Fuchsias

FUCHSIAS

*A Complete Guide
to their
Propagation and Cultivation
for House and Garden*

STANLEY J. WILSON

FABER AND FABER
3 Queen Square
London

First published in 1965
This New and Revised Edition published 1974
New and Revised Edition published 1974
this Paperback Edition published 1976
by Faber and Faber Limited
3 Queen Square, London W.C.1
Printed in Great Britain
by Butler & Tanner Ltd
Frome and London
All rights reserved

ISBN 0 571 11047 9

Preface to Third Edition

The fuchsia has now very definitely thrown off that 'Victorian' tag which so many people put upon it, and has established itself as a popular international greenhouse and garden shrub.

The term 'Victorian' was British and can hardly be applied to the specialist societies which abound and flourish in Canada, U.S.A., Australia, New Zealand, and Holland, or to the other lands where such societies are now being, or have been formed.

The joy of growing these flowers seems even to have penetrated the Iron Curtain, for a book devoted entirely to the fuchsia has been published in Czechoslovakia, and the varieties it recommends are those shown in this book.

Since the second edition of this work was published the cultivation of the flower has not altered. Better fuchsias are definitely being grown and this is due to the experience gained by the enthusiast rather than to any secret formulae.

In this third edition it will be noticed that not so many of the cultivars have been discarded as between the first and second edition. Not only is it that the cultivation is the same, but growers now recognize a good fuchsia when they see it, and only the best of the new varieties survive in the commercial catalogues beyond a year or two. However, changes have been made, and those cultivars that have already proved themselves have taken the place of those which, if not already vanished, are only lingering in individual collections.

Chapter XXII's 'Recent Introductions' has a list of cultivars which have yet to prove their worth. Some may well make the grade and become popular, but as growers are becoming more choosy, it is evident that a number of them will be almost unknown in a few years' time.

I am grateful to Ernest Crowson, F.R.P.S., A.I.I.P., who supplied eight new black and white photographs, for this edition.

7

Preface to Second Edition

Since the first edition of this work on fuchsias was published many hundreds of new varieties, or cultivars, have been introduced to the ever enlarging and hungry market. Many of the newly introduced fuchsias will inevitably 'fall by the wayside' and may well eventually be lost completely to cultivation. Among these will be those of poor constitution, and those which have nothing extra to offer on the already existing favourites. The most unfortunate casualties are those which are different and of good constitution, but which somehow fail to catch the imagination of a fickle public, and therefore become lost because they are of no economic value to the nurseryman.

Many of the fuchsias listed in the first edition have now been dropped from the current catalogues of the specialist nurserymen for the reasons already given, but while those of poor constitution, and those guilty of 'sameness' may be lost for all time, many of those which have been deleted from the list because they are unfashionable may well return as future favourites, as indeed many of present-day 'old-fashioned' varieties have.

The revised chapters in this edition have been altered to include the varieties listed in the leading specialist catalogues. In addition Chapter XXII's 'Recent Introduction' has been completely revised to include not only very recent raisings, but varieties which have only recently been introduced into Great Britain and have yet to be proved in their new environment.

Not all varieties introduced since the first edition was printed have been introduced into this edition as space does not allow this, but it is felt that all those of some significance have been included.

Preface to First Edition

No specialized work on any aspect of gardening, whether it be fuchsias or any other subject, can truthfully be described as the work of a single person. The science of horticulture has developed over centuries and the momentum of that progress gains apace each year, with every gardener by his, or her, experiences contributing a little toward the enjoyment and pleasures of the gardeners still to come.

I express sincere thanks to my many friends in the British Fuchsia Society who have given so freely of all the information that was asked of them. Particular thanks are due to Bernard W. Rawlins, current President of the B.F.S., and a leading specialist nurseryman, and to James Travis, nurseryman, hybridizer, and collector of fuchsias, for their unstinting aid on subjects that might have become difficult but for their help.

Further thanks are due to Charles W. Unwin, Histon, and Mr. and Mrs. H. F. Barnes, Cheam, Surrey, all members of the B.F.S., who first suggested that I put on paper the knowledge I have gained by growing and studying the fuchsia over many years.

My gratitude is extended to the publishers for their patience and understanding, and to Lawrence D. Hills, whose suggestions as to how the job should be tackled are reflected throughout this work.

For the photographs which are an important part of this work I thank J. E. Downward, F.I.B.P., and R. O. Couchman, for the skilful way in which they have applied their art in reproducing such detail in the subjects they have photographed.

To my wife, Joyce, who beside the household chores, and the care of three children, somehow found the time to type the book from a muddled scrawl of bad handwriting I give my most grateful thanks. Her advice as to how certain passages might

9

read to assist the beginner and small grower, without insulting the expert, did much to prevent the book becoming too technical, a fault, it seems, with authors who run out of ideas.

I am particularly proud of the line drawings, the work of my 15-year-old son, Stephen, who although without a pattern to copy was able to produce them from my vague verbal instructions.

There may well be throughout this work ideas which many an acknowledged grower will disagree with. Such points of controversy which stimulate constructive argument are normally good for the furtherance of a subject, and if this book, besides helping the grower to a better understanding of fuchsias, encourages original thought and ideas, it will have achieved a double purpose.

Contents

Illustrations

CHAPTER I

The Fuchsia and its History

It is sad that in history the fuchsia was not known in time for the beauty of form and colouring for which it is noted to be emblazoned on pennants, shields and banners of the great kings and their knights in medieval times. Had it been a plant of the Old World, when the use of purple was restricted to royalty, there is little doubt that the fuchsia with its intense colours would have been a royal flower, and that at some time in history there might well have been a 'Battle of the Fuchsias' just as there were 'Wars of the Roses'.

The fuchsia, however, had a history of its own, dating well before our own medieval period and long before it was discovered by the Western World in 1703. The rapid rise to popularity that followed the introduction of the flower to the public, its decline after the outbreak of the First World War, and the subsequent steady climb to become once again a flower for every garden and greenhouse, is modern history; and the fuchsia had had a long history centuries before these events took place.

A great number of the species of fuchsia originate in countries that were many years ago the empires of two great civilizations, the Aztec of Mexico and the Inca of Peru. From records of Aztec history it would appear that they little cared about the existence of the fuchsia, for it grew in difficult mountain terrain in areas inhabited by insignificant tribes of Indians. Very little is known of the details of Inca history, for this nation never discovered how to write, and the string records (known as 'quipus') that it kept were practically all destroyed by the Spanish conquerors. They did, however, build their cities high in the Andes where the fuchsias grew, and there is every reason to believe

15

that they enjoyed its presence. The beautiful species *F. Sanctea-Rosea* was actually discovered in 1898 growing in the ruins of that famous Inca city, Macchu Picchu, and it may well have been the descendant of a plant that once adorned the house of an Inca nobleman.

The Incas were not the first civilization to inhabit that part of South America where so many of the fuchsia species abound. They were, in the main, agriculturists rather than horti-culturists, and it is to them that we owe such vegetables as the potato, tomato, bean, both 'French' and runner, maize and the sweet potato. The 'Garden of the Sun' of their last ruler Atahualpa, which had all these vegetables and fruits fashioned in pure gold, was lost when the galleon carrying the treasure to Spain was lost in a storm.

It was probably an earlier civilization, the 'Chimu' (A.D. 1000), a very artistic people who became vassals of the Inca, who became sufficiently interested in the fuchsia to draw its likeness with remarkable accuracy on cave walls and cliff facings. Although the identity of the artists has never been really confirmed, the experts are all agreed that only craftsmen of a civilized community, equipped with well-made tools, could have cut such figures into the hard rockfaces.

With the collapse of the Inca civilization, the subjected tribes of the northern Andes, who for centuries had tilled vast acres of agricultural land for their rulers, were scattered far and wide, and it is known that many owed their lives to the fuchsia berries on which they lived, while fleeing from the tyranny of their new Spanish overlords. To this day many tribes of Indians relish the flavour of the fuchsia seed berries, which they gather from the wild plants in much the same manner as we gather our native blackberry. The fruits of the species *F. fulgens*, with its green berries, and *F. corymbiflora* and *F. boliviana*, with their large juicy purple fruits, have extremely pleasant flavours besides being rich in vitamin content. The berry of *F. corymbiflora* has been described as having 'a flavour resembling that of a well-ripened fig'.

Besides those species found in Central and South America, there are six which are found many thousands of miles away on the islands of New Zealand. Historians and scientists have long held the theory that islands such as Tahiti and New Zealand were populated by people who originally came from the coastal regions of South America. Not only do the people show similar characteristics, but much of the flora has a similarity which proves conclusively that there is some connecting link between these islands and South America.

This link is undoubtedly the trade winds and the South Equatorial Current which carried the 'Kon-Tiki' Expedition on their balsa-wood raft from the coast of Peru to the islands of Polynesia. The ancient ancestors of the Maoris, and the Polynesian Islanders, who originally made this journey centuries before, could well have carried large quantities of dried fruit to supplement the fish diet they would have been obliged to live on. Among the fruit may have been fuchsia berries, and from the seed came a new race of fuchsias.

The New Zealand species are very different from those of South America, but they developed under different conditions, as did much other flora which was carried in this way, and Nature made adjustments to suit such conditions.

The species *F. excorticata* grows to tree-like proportions in its native habitat and is one of the most common trees of the New Zealand bush, where it is easily recognized by the characteristic cinnamon-brown bark that hangs in long strips. Here again the fuchsia had a history long before Europe knew of its existence. The masses of bright blue pollen for which *F. excorticata* is noted was used by the Maori maidens, long before the arrival of Captain Cook, as a face powder. As the Maoris based a great deal of their lore on what Nature had put around them, it is thought that the idea of powdering the faces blue was in imitation of the nectar-sucking birds which inhabit the bush in vast numbers, and who get their faces smothered in the blue pollen as they partake of the nectar. This tree species was affectionately known by the Maoris as 'Kotuhutuhu' and the berries, which are still enjoyed by the children even to this day, were known as 'Konini', a name which has now been accepted by the white population of New Zealand for general use. The Maori tribal chiefs had still another use for this species. About a century ago, after the signing of the Treaty of Waitangi, whereby the Maoris accepted British rule, and native executions were forbidden, it was a profitable sideline to tattoo the heads of slaves elaborately, decapitate them, and cure the heads in the juice of 'Konini'

17

mixed with that of 'Bidi-Bidi', *Acaena sanguisorbae*. This gave the decapitated head an antique appearance and the souvenir-hunters of that day were fooled into paying high prices for these gruesome relics.

Although the fuchsia does, to many people, give the garden an 'old-world' air, its history as a garden plant or shrub does not go back very deeply into time. In 1703 emigration to the New World was being encouraged by many governments who wished to share in the wealth that was there, but their progress was impaired by the terrible scourge of malaria. To combat this malady many eminent botanists were sent to find more sources of cinchona trees, from which quinine is extracted. One such botanist was a missionary and devout member of the religious order of Minims, Father Charles Plumier; during his travels he collected and made drawings of a number of plants, nearly all of which he named after eminent botanists who had preceded him in history, such as Turner, Gerard, and Parkinson. While searching in the foothills of San Domingo, he made his greatest find, which he called *Fuchsia triphylla flore coccinea*. He named this very beautiful genus, Fuchsia, after Leonhart Fuchs, who was born in 1501 and for thirty years until his death in 1566 occupied the chair of medicine at Tübingen, one of the great theological universities. Fuchs's work *De Historia Stirpium*, which was published in 1542, is still regarded as a work of reference by herbalists, and there is no doubt that the studies of Father Plumier were greatly influenced by his teachings.

Plumier published his work *Nova Plantarum Americanarum Genera* in 1703, in which he described, among other plants, his discovery of the genus *Fuchsia*. The actual drawings of the fuchsia in this work were very crude and inexact but there is now no doubt that Plumier's flower was the *F. triphylla* that we know today. Linnaeus in his first edition of *Species Plantarum*, in accordance with his rule of binominal nomenclature dropped the '*flore coccinea*' and this, the first recorded species of the genus *Fuchsia*, became just *Fuchsia triphylla*.

Although Linnaeus accepted the drawings of Plumier as fact, their inexactitudes created confusion to both botanists and gardeners for nearly two hundred years after the discovery. No further plants were found and many began to doubt its exis-tence, although by now many other species of fuchsia were in general cultivation. It was in 1873, when an American, Thomas Hogg of New York, sent home seed he had collected in San Domingo, that the botanists were able to confirm that *F. triphylla* existed and grew naturally in the West Indies.

The introduction of the first fuchsia species to Great Britain will always be shrouded in mystery. It appears in the first place that a Captain Firth presented Kew Gardens with a plant of the Brazilian species, *F. coccinea*, in 1788. During this same year the same species was somehow acquired by James Lee, an eminent botanist and nurseryman of Hammersmith. The story is told that James Lee was informed of a strange plant that had been seen growing in a window box in the back streets of the district of Wapping, London. As an astute businessman he immediately made inquiries for himself, and found that the plant was owned by a widow whose son, a sailor, had given it to her on his return from a South American trip. He somehow induced the widow to part with her new plant for the sum of 8 guineas, and immediately set about building a stock of what is said to have been three hundred plants. There is no real evidence to substantiate Lee's story, but he was undoubtedly a great man of his day, and as there is likewise no evidence to disprove his story it would be honourable to say that perhaps both the Kew plant and the widow's plant were introduced into this country together. There is no record as to who the sailor was, but he may well have been a member of the crew of the ship of which Captain Firth was in command. It is hardly likely that somebody as knowledgeable as Captain Firth appeared to be, would have been content to bring only one plant with him, knowing that it had to face a long and arduous journey, during which time it might well succumb. It could have been that he started off on his journey with several plants of fuchsia, some of which died *en route*, and one was lost to a crew member with a distinct eye for beauty.

Certain facts were however recorded, and they were that the first fuchsias were sold to the public in 1793, at prices ranging from 10 to 20 guineas each according to the size of plants, which means that James Lee made an extremely handsome profit. The plant-hunting expeditions which were now being sponsored by the larger nurseries and other organizations were quick to realize that the introduction of new fuchsia species was both necessary and profitable and in 1796 *F. lycioides* was introduced. This was followed by *F. arborescens* (1824), *F. magellanica* (1827), *F. fulgens* (1830), *F. corymbiflora* (1840), and *F. apetala, F. decussata* and *F. serratifolia* in 1843 and 1844.

This wealth of new introductions soon brought about the inevitable demand from an insatiable gardening public for something new, and something different, and so the hybridizer came into his own. The first recorded attempt at hybridizing is believed to have been made in England, and was a crossing of *F. arborescens* × *F. coccinea*. The result of this crossing is not known, but a start had now been made.

The first garden varieties were variants of the species *F. magellanica*, of which there were quite a number, and because of the versatility of this species the early hybridizers concentrated on using it as a parent. Records show that early crosses were mainly between *F. fulgens* and *F. cordifolia* together with *F. magellanica*. From this hybridizing came many interesting and fascinating plants, but the continual use of *F. magellanica* has proved in later years to be rather a handicap, as crosses made between choice varieties often produce, in the resultant seedlings, the dominant features of this species.

A real breakthrough came in 1842 when the variety 'Venus Victrix' was sent out by Mr. Cripps, a nurseryman of Tunbridge Wells, who sold the plants at 1 guinea each, and described it in his catalogue as follows: 'The flowers of this unique variety are white sepals delicately tipped with green, with a superb bright purple corolla.' This variety was actually an accidental seedling raised by a Mr. Gulliver, gardener to the Rev. S. Marriot, of Horsmonden. As a plant it was quite diminutive and could not by today's standards be regarded as anything special, but it was the first real variety to have an almost white tube and sepals; and, more important still, it proved to be an outstanding parent, imparting this same characteristic to many of its progeny. For many years after its introduction this variety was used to give a wealth of other varieties, both directly and indirectly, bearing the white-tube-and-sepals hall-mark.

It was in 1850 that the next major step in fuchsia hybridization was made. Mr. Story, a nurseryman of Newton Abbot, raised in that year the first double-flowered hybrids. There was evidently nothing outstanding in these first double varieties, but progress had been made, and the hybridizers were quick to improve on them. Mr. Story also made further progress by introducing, also at this time, the first of the white-corolla varieties, both single and double. The most famous of these varieties was 'Queen Victoria', named after the reigning monarch.

It is interesting at this stage to reminisce on the progress made up to this time. There were now on sale to the public varieties which had the white tube and sepals with coloured corollas, and other varieties which had coloured tube and sepals, but with the white corolla. To anybody not understanding fully the parentage of these varieties, it would seem perfectly simple to cross a variety with the white tube and sepals, with another variety with the all-white corolla and produce from the resultant seedlings a perfectly all-white flower. This is, however, where the almost indiscriminate use of *F. magellanica* as a parent, in the early days of hybridizing, began to have its effect. It was a case of 'the sins of the father falling upon the third or fourth generation', and all the brains and experience of the then extensive fuchsia world were unable to unravel this puzzle. A hundred years were to elapse before the American plant breeders of California, with their modern appliances and their knowledge of the use of drugs in hybridizing, together with a system of inbreeding, outbreeding, and continuous selection and re-selection, were to market the first almost pure white fuchsia.

Fuchsia growing was now an art in itself and the plant's popularity was not confined to England. In France in 1848 M. Felix Porcher had published a second edition of his book *Le Fuchsia, son Histoire et sa Culture* in which he describes 520 species and varieties. Unfortunately a large number of the varieties listed were merely duplications of each other, for introductions from England were given French names, and varieties raised in France were given popular English names in this country to facilitate their sale. The confusion caused by this practice is with us to this day, although the efforts of the British Fuchsia Society to clear up the matter have done much to put things right. However, the publication of M. Porcher's book was a notable happening in fuchsia history, for despite its inaccuracies it was the first serious attempt to record the progress and hybridization of the plant. This book was

21

The Fuchsia and its History

eventually to have four editions, and it is noticeable that with every edition the author became more and more knowledgeable.

It was also in 1850 that Mr. Story sent out the first fuchsia with a striped corolla. Another eminent hybridizer and nurseryman, Mr. W. Bull of Chelsea, followed this with a variety *F. striata* var. 'Perfecta' in 1868, and a double striped variety, 'Gipsy Queen', a year later. None of these varieties was able to stay the course and the only really old variety which can be described as striped, and which is still available to this day, is 'Bland's New Striped' which was introduced by Mr. E. Bland in 1872. Those worthy varieties having multi-coloured corollas, which are so popular today, are mainly modern introductions from the plant breeders in America.

The next fifty years in fuchsia history is a story of increasing popularity, with this beautiful flower firmly endeared to our Victorian ancestors. New varieties flooded a ready market, and new raisers were ever ready to give their customers new introductions, some of which were good and survive to this day, others bad or indifferent and quickly rejected, to pass out of cultivation. Fuchsia growers will be ever thankful to British raisers of such standing as F. B. Salter (1837), Standish (1840), Story (1842), W. Bull (1846), E. Bland (1847), Lye (1862) and Veitch (1894) and their continental counterparts: Lemoine (1848), Cornelissen (1857) and Rozain-Boucharlat (1860). The dates given are those of their first introductions. There were many more names than this but for both quantity and quality these hybridizers stand out above all others.

In 1883 the first book written in English on fuchsia cultivation was published. This book dealt mainly with cultivation, together with details of how to grow and train pyramids, and other popular shapes of that day, and was entitled *A Practical Treatise on Fuchsias*, by Frederick Buss.

The late nineteenth century was indeed a great gardening era and it must be recorded that the fuchsias at this time were grown to a size and perfection, both for the exhibition hall and for the huge garden parties that were the rage at that time, the likeness of which has never been seen since. Photographs show pyramid-trained plants growing fully 10 feet high, and 5 feet through at the base, and a mass of flower from base to apex. This day of the large gardening staff and large conservatories suited the growing of these amazing giants very well, and it was often the case that one member of the gardening staff had nothing more to do than be responsible to the head gardener for the fuchsias alone. It may well be that with the 'hurly-burly' of the present day, coupled with the dwindling size of gardens and the acute shortage of good gardening personnel, we may never see such beautiful creations again, unless automation eventually does all our work for us and we have nothing left to do but help the fuchsias to grow.

No history of the fuchsia would be complete without mention of James Lye, the man who was perhaps more than anybody else responsible for exhibiting some of these remarkable plants and bringing them to the notice of the public. Born in 1830 in Wiltshire, he spent the whole of his gardening career in the service of the Hon. Mrs. Hay, a sister of the Earl of Radnor, at Clyffe Hall, Market Lavington, in Wiltshire, where he rose to the exalted position of head gardener. Apart from the wonderful fuchsias he exhibited, he was a prolific hybridizer of the shrub, and many of his excellent introductions are fortunately still with us. He was particularly noted for the varieties that had cream tubes and sepals, and this combination eventually became known as 'Lye's Hallmark'.

The next step in the breeding of new fuchsias was not really taken until the turn of the twentieth century, and this was not so much an advancement as a new shape and colour-break in the flower. From Germany came a new race of fuchsias which are now termed the 'Triphylla Hybrids'. The parentage of these hybrids was kept a dark secret by their raisers, Rehnelt and Bonstedt, but by their habit of growth, the foliage and the flowers it is practically certain that most of them are mainly, if not solely, the results of crosses between the species *F. triphylla* and *F. fulgens*. As stated already they have given a new shape to the hybrids, but they are unfortunately (like their parent *F. triphylla*), extremely frost-shy and generally more tender than most fuchsias, and can only survive an English winter with the aid of artificial heat.

So in the early part of the present century the fuchsia continued to increase in popularity. The marvellous exhibits of fuchsias that James Lye had given to the horticultural world were now emulated by his son-in-law, Mr. G. Bright of Reading.

Gardeners generally had the fuchsia very much in mind, and many hundreds of thousands of plants each year, many of them varieties still popular to this day, passed through the market and into the homes of ordinary people, where they contributed to the gaiety of this already gay and colourful era.

The dark days of 1914 saw the start of the decline in popularity of the fuchsia, a decline which was almost to obliterate it as a specialist flower. It is understandable that during the war years with the continual 'Dig for Victory' campaigns, all the efforts of gardeners and nurserymen were turned towards the growing of food for the nation, and that flowers of all kinds would suffer in consequence; but what is not so understandable is that when peace returned the popularity of the fuchsia did not.

There are two schools of thought among fuchsia growers as to why the decline continued when peace returned. The first is that the transition from uniform to civilian dress created a completely new thinking public, a public in constant rebellion against the past and all that it stood for. The fuchsia was thought to be a symbol of that past, and was therefore rejected along with all the other things considered 'Victorian'. The second school of thought reject this explanation and state quite emphatically that the rebellion was only against the bad things of the past and certainly the fuchsia could not be considered in this class. Their explanation of the decline is very logical and it is this. The 'Dig for Victory' so necessary during the war years brought to the notice of the general public, for the first time, the delights of the tomato. Practically every greenhouse was turned over to its cultivation and it is most unfortunate that the fuchsia and the tomato have only one thing in common with each other, and that is the notorious greenhouse pest white fly (*Trialeurodes vaporariorum*). Modern science has ensured with its chemicals that the white fly need no longer be considered a serious menace, but during those days in question there was little that could be done to check the spread of the pest. In actual fact tomatoes are an equally good host, and a study of the pest records that they breed as happily and as quickly on tomatoes as they ever did on the fuchsia.

Whatever the cause of the decline in popularity, these were indeed sad times for a flower that had been held so high in the public esteem for so long; but tucked away in many a cottage garden and many an odd corner of a greenhouse, plants continued to grow that were to become the nucleus of the new fuchsia age.

Although in those lean years of the fuchsia's garden history Great Britain could still boast of specialist fuchsia nurseries, and a wealth of varieties scattered unnoticed throughout the land, the impetus which was to start the fuchsia on its road back to favour came, not from this country, but from the United States of America. The formation in the state of California of the American Fuchsia Society in the year 1929, not only ensured the future of the flower, but started a wave of enthusiasm which was to spread to many parts of the world. This Society saw the enormous potential that the flower possessed and in 1930 sent to Europe three representatives, Miss Alice Eastwood, Mrs. Evelyn Steele Little and Professor Sydney B. Mitchell of the University of California. Their aim was to visit nurseries and private and botanical gardens and to collect data and specimens of the varieties that they thought would suit the climate of California, and which could be further developed by the hybridizer. From a nursery in England they collected fifty varieties of fuchsia and from France a similar number, and it was from this beginning that was to come the greatest hybridizing spree that the fuchsia world had seen since the last century. Fuchsias of different shape and colour, tiny flowers and new giants, began to pour out of California; and the names of the hybridizers, V. Reiter, Niederholzer, Tiret, Hazard and many others were coupled with the names of a whole host of new varieties.

In the year 1936 Dr. E. O. Essig, Professor of Entomology at the University of California, made a brave attempt to sort out with his book *Check List of Fuchsias* the confusion which existed over the nomenclature of the varieties of fuchsia known to exist. Up to this date his was perhaps the most outstanding publication ever to be devoted entirely to this shrub, for he listed over 1,900 different names; and although it has by now naturally outdated itself it is still used as a basis for up-to-date check lists.

The next major move in history was to come from London. A group of enthusiasts met in 1938 and the Fuchsia Society, later to be known as the British Fuchsia Society, was formed. In the first year of its formation it issued a year-book which was the first publication dealing with fuchsia culture written in the English language since the turn of the century. The original list of founder members, which was published with this book, read almost like a page' from Debrett; and so great was the enthusiasm of these members that not even the chaos of the Second World War, which hit them almost as soon as the Society was formed, could deter them from their dedicated task of bringing the fuchsia back to public favour, in the country that had first adopted it as a garden shrub. That the work of the Society was able to continue during the war was mainly due to the efforts of their secretary, Mr. W. W. Whiteman, who continued to publish a year-book on behalf of the Society, and acted as a liaison officer between the members, as they were scattered throughout the world by the various duties imposed on them by the emergency. Throughout this period the fuchsias of the country cottage garden and those kept in odd corners of many greenhouses were hunted, identified, propagated, and distributed by the Society, and other enthusiasts set about the task of hybridizing so that something new could be offered to a fickle public.

The Second World War over, the impetus gained momentum despite the untimely death of Mr. Whiteman in 1945. New varieties flowed freely from America, and from that same country Dr. Philip A. Munz made in 1944 the first serious scientific attempt to sort out the known species in the genus *Fuchsia*. His work *A Revision of the Genus Fuchsia* should be taken into account by all those who wish to make the fuchsia a subject of scientific study.

Not all things new were to come from America. The British Fuchsia Society had a nucleus of enthusiasts who were determined to bring the flower of their choice to the fore. One of these members, Mr. W. P. Wood, a retired professional gardener who was himself a collector and a great exhibitor of the flower, felt that only when the fuchsia was relieved of the tag 'greenhouse shrub' would the general gardening public take the genus to its heart. With this object in view he set about breeding a new race of hardy fuchsias. Up to this time the hardy varieties were generally small and of very limited colour range, but Mr.

Wood made great strides, not only in increasing the size, but in introducing new colours into the hardies, in many ways equal to those of varieties considered to be tender. Furthermore, he set out to prove that many of the existing varieties which were catalogued under the heading 'Greenhouse Varieties', were in fact capable of growing happily in the shrub border, and were hardy enough to stand the rigours of an English winter in most parts of the country. Unfortunately his work was only half completed when he died suddenly in 1951. However, a great deal of his early work was recorded in his book *A Fuchsia Survey*, published in 1950.

This now brings us to the present day and the future. What are the prospects of the fuchsia in the days to come? In America two large societies, the American Fuchsia Society and the Californian Fuchsia Society, together with their smaller neighbour the Oregon Fuchsia Society, continue to flourish and wax strong. In New Zealand, the New Zealand Fuchsia Society grows larger and healthier each year. At home, the British Fuchsia Society increases its membership yearly to such an extent that its amateur administration finds it more and more difficult to cope with the work thrown upon it, and throughout the country enthusiasts are forming themselves into local groups and societies devoted entirely to the cultivation of their favourite flower. New varieties continue to be introduced each year, and hybridizers both in America and Great Britain are now working with definite objects in view, new colours, stronger plants, and larger flowers in greater numbers. All these efforts must be capped with success and there is little doubt that more and more gardeners will be captured by the charm, grace, simplicity, and ease of growth of fuchsias and will undertake their cultivation.

To name and record the efforts of everyone who has helped to give the fuchsia a place in garden history would in itself require several volumes, but among the many varieties that have proved themselves outstanding, and the newer varieties which have given every indication of being potentially great and which are listed in Chapters XII–XXII of this book, will be found the names of many of the 'greats' in fuchsia history, past and present, whose names will ever be commemorated by the varieties they have raised, or by the varieties named after them.

CHAPTER II

Fuchsia Hedges

In these days of rapid travel, more and more people who know little about gardening, and even less about the fuchsia as a flower, are being impressed by the wonderful sight of fuchsias in full bloom that they meet on their holidays in the south-west of England, the west coast of Scotland, and Co. Kerry in Eire. It is in these places particularly that the hardy fuchsias have naturalized themselves, and during the peak of the English holiday season are a dominant feature of the landscape. Whole roads and country lanes are lined with banks of red and purple flowers, and from early July to the middle of September this mass of colour persists.

There is no doubt that these hardy fuchsias, consisting in the main of the species *F. magellanica* and its subspecies *F. m. gracilis*, *F. m. molinae*, together with its hybrids F. Riccartonii and F. Thompsonii, did much to keep the name of fuchsia alive during the serious decline in popularity between the two World Wars. With the wonderful display they put up every year it was obvious that eventually somebody with an eye for beauty of colour and form would at last see the enormous potential of the flower as a cultivated feature. It is a fact that the greater number of the larger-flowered hybrids, which were hunted by enthusiasts immediately after the birth of the British Fuchsia Society in 1938, were found in the cottage and country gardens situated in those areas where the 'wild' fuchsias, as they were known, had naturalized themselves. To those cottagers and gardeners this flower was a part of their everyday life. It grew all around them and it was therefore only natural that its more glamorous hybrids should adorn their gardens.

How unfortunate it is that while the popularity of the fuchsia

28

continues to increase rapidly, the enormous potential of the hardy species and varieties as hedging plants has never been realized to the extent that it deserves. Evidence points to the fact that this reluctance to use the flower for this purpose is due more to an ignorance of the flower itself, and its capabilities, than to an actual dislike of the plant.

Many people, with many keen gardeners among them, are convinced that the fuchsia is a plant of the West Country and is not suitable for planting elsewhere. This is a complete fallacy, for dotted at random throughout the country can be seen individual specimen bushes, some of considerable size. Where one plant grows there is no reason to suppose that dozens will not grow. The luxuriance of the growth may vary from place to place according to the conditions, both geographical and climatic, but provided the hardy fuchsia—whether species or variety—is not planted in a frost pocket, and its meagre requirements of cultivation are met, it is certain to grow and flourish.

A vast sum of money is spent annually by the holiday resorts, both large and small, with a view to enticing the ever-increasing holiday crowds to visit them. Much of this money goes on supplying flowers and plants for the beds on the seafronts and elsewhere in the towns. What is surprising is that very little effort is made by the authorities to beautify the approaches to the resorts. So often are the outskirts of the towns heralded by the sight of an untidy caravan- or camping-site, the town gasworks, or a corporation rubbish dump. First impressions of anything can be most important, and how much more pleasant it is to enter a town by driving, walking, or riding along a road coloured by these beautiful shrubs. It is not suggested that the fuchsias themselves will hide the eyesores that civilization has imposed on the countryside. This is the job for the poplars and the forest giants, but the beauty of the hedges will do much to distract the eyes from less beautiful things.

In the fuchsia we have a subject that is ready-made for the production of low hedges, besides being the one shrub that will bloom continually throughout the whole of the English holiday period. The majority of shrubs that flower during the months of July to September produce bloom for a comparatively short period and leave behind them unsightly seed-pods and dead flowers. Many may argue that the rose will flower continuously

29

during this time, but the dramatic spread of 'black spot' disease in the country areas, and the crippling effect of insect pests on the more glamorous members of the rose family, means that to get full value from these otherwise delightful subjects a greater demand will be made on labour and money, two items which are hard to come by for any public authority.

Gardeners themselves can be criticized for their neglect in using the fuchsia as a hedging plant. Only comparatively few gardeners have used this shrub as a low hedge dividing one section of the garden from another. In some of the large gardens which are open to the public at various times of the year, these hedges have become quite a feature in themselves besides being ideal for their purpose. If any criticism is to be made concerning those gardeners who have used the fuchsia for the purpose of hedging, it is their lack of adventure in their choice of variety used. There is no doubt that the majority of gardeners are under a very false impression that the hardy fuchsias are only of the 'red and blue' type. This notion is perhaps understandable as practically all the species and the varieties that have naturalized themselves are of this colour combination. There are, however, a large number of different colours within the range of well-tried hardy fuchsias, and the reader will do well to study the list of varieties suitable for hedges that is given in Chapter XIII at the end of this book.

A brief study of the geographical and climatic features of the district where the fuchsia has naturalized itself may help the reader to understand some of the requirements of the shrub, and so assist the establishment of fuchsia hedges elsewhere. While it is not possible to coax the Gulf Stream, or more correctly the North Atlantic Drift, to prevail on all our shores, we can understand the conditions that it creates, for quite obviously these are the conditions enjoyed by the plant. The part of the Gulf Stream known as the North Atlantic Drift strikes the whole of the west coast of the British Isles from Cornwall in the south to the Orkney Islands in the north. As this warmer current meets the colder waters of the Irish Sea, St. George's Channel, the English Channel, and the North Sea, so moisture is given off into the atmosphere and this moisture-laden air is carried inland by the prevailing westerly and south-westerly winds. It is this cool moist air that makes the fuchsia grow so luxuriantly in certain parts of these islands. It is unfortunate that so little of this warmer air ever reaches the eastern coasts of England and Scotland, which parts furthermore have to contend with the dry and often bitterly cold winds that sweep across Europe from Siberia.

The ideal conditions which induce the fuchsia to give of its best are therefore damp and cool surroundings. The hot and dry, and the cold and dry, atmosphere call for a special survey of the planting positions in order to get the best results from the plants. Perhaps the greatest enemy a fuchsia can have is a dry wind, whether it be hot or freezing cold. A hot dry wind will drain the succulence from the leaves and cause serious scorching, and the cold wind will cause the leaves to respire too rapidly, leaving them unprotected from the extreme cold. While it is impossible for anyone to supply the winds with the necessary moisture, it is possible to supply windbreaks which will help to keep these killers from the hedges themselves, or at least break the ferocity of the blasts enough for the fuchsias to have a chance to establish themselves and thereafter flourish. It would be foolish to suggest that these windbreaks be erected merely for the purpose of establishing a fuchsia hedge. It would instead be common sense to realize that any garden subject to the conditions described must have some form of protection for all the subjects being grown therein.

Nature itself provides shrubs and trees which form the most effective windbreaks. Only the garden of estate proportions can consider the large trees as a shield against the wind. The smaller garden will probably have to resort to shrubs such as privet, the hedging honeysuckles, *Lonicera nitida*, *Tamarix*, or tree hedges of beech and hornbeam, yew, holly, or Lawson's cypress. For the very small garden where space is so precious, perhaps the ideal would be the wattle hurdles that can be purchased from most horticultural sundriesmen. Whichever method is used for the purpose of protecting the garden from the wind, great care must be taken to prevent the formation of artificial frost pockets.

It is well known that plants on the side of a hill will often escape frost damage, while those in a valley are killed by the effect of the frost. This is a result of the flow of cold air down-hill into the valley. The air cools and becomes denser and heavier

31

after nightfall, and the air in the valley, since it is already as low as it can get, remains in position and becomes colder and colder. Windbreaks wrongly sited can check this flow of cold air downhill and form an artificial pocket. Only subjects of extreme hardiness can withstand the conditions described, and while the hardiest of fuchsias are known to survive such conditions, one can hardly say that they are conducive to the growing of a flourishing hedge, which is what one should aim for.

Many of the newer housing estates are being built without the traditional front brick wall or garden fence. The growing of privet and other close-knitted shrubs spoils the open effects that the planner tries to create. If a hedge is at all necessary it would be so much better to have a fuchsia hedge, which by nature of its growth is light and open, is easy to control, can be grown in rows of different colours, and does not destroy the wide-open-space effect that the twentieth-century planners strive for.

What of the cultivation necessary for the growing of the hedge? It must first be remembered that a hedge must always be thought of from a long-term angle. Once it is established it is expected to grow and flourish year after year with the minimum of attention. With this in mind it is not too much to expect that the maximum attention be given to preparation before planting; and when planted the young shrubs must be carefully tended to ensure their firm establishment.

Fuchsias by their nature demand a deep cool root run. This means that the first essential part of the preparation is deep digging. The object of this operation, for those who feel that deep digging is a waste of time, is to break up the subsoil, and at the same time add to it material which will prevent it packing down again before the fuchsia roots have established themselves at a depth. On the other hand some subsoils are so loose that material may have to be added to bind them and prevent the loss of too much moisture.

When you have dug at least two spits deep the subsoil should be forked over; and according to the nature of the subsoil, binding or loosening material is added to it. In the case of a very loose subsoil such as sand or gravel, the addition of peat, undecayed leaves or any bulky water-retaining material of that nature should be ideal for the purpose. The same treatment would be ideal for a chalky subsoil, as the acid nature of the material would help balance the high alkaline value of the chalk until the hedge was established. With clay, of course, the material added must be light and of open texture. In a very heavy clay it would be advisable to incorporate gravel or very coarse sand to prevent waterlogging of the young roots. It has been known for people to collect large quantities of dried twigs from the woods and work them into clay to help keep it open.

Farmyard manure has not been mentioned as one of the materials to bury beneath the second spit. Gardening books have for generations advocated the placing of farmyard manure beneath the plants for the roots to work down to, but its general shortage makes it a material far too precious to bury. Furthermore, science has proved to us that the feeding roots of a plant, shrub or tree are situated just below the surface of the soil. The deeper-penetrating roots are there to provide an anchor for the subject and to forage for moisture. It is this moisture that is so essential to the young fuchsia at this stage. The fuchsia has a succulent leaf which has a high rate of respiration and therefore requires a constant supply of moisture, but any waterlogging of the subsoil will mean the exclusion of oxygen so necessary to healthy root formation, with the subsequent dying of the root hairs, resulting in sickly, struggling plants.

The subsoil having been dug over, it is now important that the second spit be returned first and that the top spit should remain on top. So many people make the mistake of bringing the second spit to the surface and burying the all-important top layer of soil. As already mentioned, the feeding roots of the fuchsia will be just below the soil level, and as all the weathering, all the bacterial activity, and the natural humus are in the top few inches of soil, it is obvious that any important upset of this sequence will affect the quick get-away and ultimate establishment of the hedge.

It is better if this initial preparation of the soil be done during late autumn or early winter so as to allow the soil to weather and settle well before planting time. Bulky, moisture-retaining materials should be incorporated with the soil as it is returned. Horticultural peat, spent hops, and coarse leaf-mould are ideal for this purpose. Fertilizers should not be added at this stage as much of them would be leached from the soil by the winter rains. In any case, food and nutrients are only welcomed by plants that have the capacity to absorb and benefit from them, and can be added when the hedge is growing strongly.

It is assumed that the site of the hedge or hedges is now prepared, so the next consideration is the fuchsias themselves. Although the planting time is not until late May in the south, or early June in the north, it is grossly unfair to expect any nurseryman to have the necessary plants for when the customer wants them unless he is given plenty of notice. Most fuchsia nurserymen start their stock plants into growth either just before, or just after, Christmas so that by the time the spring trade gets under way they have an ample stock of plants to meet the demand. Therefore to be reasonable and fair you should give the nurseryman the order to put on his books before Christmas. This is the only way to avoid disappointment.

If it is the intention to grow a fuchsia hedge—and the result will never be regretted—let an appeal be made to the sense of adventure that is the hall-mark of a good gardener when he chooses the variety he intends to grow. It is so easy to grow *F. magellanica* and its hybrid F. Riccartonii because they are seen everywhere and the result is known before it is started, but there is a great variety of fuchsias in a large number of colours, and colour combinations, that will give a greater pride and satisfaction to the grower once the hedge is established.

The nurseryman should be advised as to when the plants would be needed. If the grower has a heated greenhouse he can take possession of the plants in early April and by his own cultivation grow the plants on to quite a large size by the time it is necessary to plant the fuchsias in the open. If only a cold frame is available to house the young plants, then late April or early May is early enough. Should the grower have no form of protection the nurseryman should be instructed to send the plants when ready to go straight into the open ground, that is, late May in the south or early June in the north, provided that the weather is normal for the time of year. The grower who intends to adopt the latter course should make sure that the nurseryman is informed of his intention, so that the plants are well hardened off at the nursery before dispatch.

The gardener who receives his plants in April must remember that they have come from a heated greenhouse, and must therefore on receipt ensure that they continue to grow in these conditions until the setback caused by travelling has passed and the plants are continuing to grow again. They should be potted up into the same size pots as when they left the nursery. This can be easily recognized by the size of the ball of soil and roots that the plants have on them when received. The whole object of the operation is to overcome the transition period with as little check to growth as possible. For those who have only the garden frame in which to grow the plants on, it will be as well to cover the frame with sacking, or similar material, for the first couple of days after receipt of the plants. This will help to minimize the vast difference between day and night temperatures that this country experiences in the spring months, and allow the plants to acclimatize themselves to their new environment.

Whatever the time of receipt, it is important for the future of the plants that are to make up the hedge that they are not allowed to become pot-bound. The plants potted in John Innes Compost No. 2 on receipt, and given an atmosphere in which they can grow, will quickly fill the pots with roots. If the plants are not potted on to larger pots before the roots become bound in their existing pots, the fuchsias will become stunted in growth and it will prove difficult for fresh root growth to break free from the original ball of soil. Such a condition, when the plants are placed in their permanent positions in the open, could well lead to the loss of some of the plants and poor growth of the remainder. This is brought about by the comparatively large area of leaf above ground keeping the moisture and rain from the small area of root. It can be likened to standing under a tree to keep dry, only in this case the dry area would be the total root area of the plants. Periodical soakings will help offset this, but there is no guarantee that even that will get past the impenetrable barrier that tightly bound roots can prove to be. Naturally the loss of any of the plants would upset the balance of the hedge, and would delay the effect that the grower is striving for, for one year or maybe two.

Before planting-out time comes, it is most essential that the fuchsias be thoroughly hardened off. There is no doubt that fuchsias, whether hardy or otherwise, moved from green-houses or frames without being hardened off, receive a shock from which they take a long time to recover. In many public gardens and parks throughout the country can be seen fuchsias, which should have dark green glossy leaves, having instead a copper-brown hue. This is a sign of insufficient hardening off, and it is easy to record the check they receive by the time it takes for the newer leaves to grow green. The *Triphylla* hybrids and the hardy fuchsia variety 'Alice Hoffman' already have the coppering naturally in their leaves, but they are the only varieties that one is likely to meet having this trend.

The professional gardener and the experienced amateur gardener have learnt the art of hardening plants off before planting out, but the not so experienced may have a little difficulty. This operation can only be generalized, as weather conditions, hours of daylight, and many other factors may make a difference to the time involved within the operation itself. If the fuchsias are moved from the greenhouse to the frame in early spring, the pots should be sunk into the soil or ashes inside the frame, the frame closed, and for the first day or so kept covered with hessian or similar material. The hessian can then be removed and the frame opened an inch or two during the hours of daylight only. The frame should be closed before darkness falls in order to trap as much as possible of the available heat inside the frame, and the hessian or covering replaced in case the temperature should drop too much. After a week, weather permitting, the cover can be left off at night, except when frost threatens, and the frame can be opened another inch or so during the daytime. After another period the frame should be opened wide during the day and left a little open at night unless frost is forecast. This goes on until the top of the frame is left off both day and night, only being replaced during extremely cold, frosty or stormy weather.

When the danger of frost is past, which in the north of England is usually early June, and in the south the end of May, the pots can be sunk into the ground in the exact position that the plants are to occupy in the garden, until the gardener is ready to plant them. Planting distances will depend on the variety being grown and the geographical position of the hedge itself. Obviously the plants will be more robust in growth in some parts of the country than in others. Depth of planting should be approximately two inches below soil level. That means that the ball of soil as grown in the pot should be that much below soil level. It will be necessary to plant the fuchsias firmly, as from now on the plants have to face the elements unaided.

No fertilizer should be given at planting time except for a dusting of bonemeal, which can be dug in when the actual operation of planting is being done.

The whole operation of planting any plant, whether it be fuchsia or not, is of the utmost importance to the future of the subject. How often is it seen that a seemingly healthy plant succumbs when eventually it is planted out. The gardener has responsibility for every aspect of the job except one—weather conditions after the planting has been accomplished. It is, however, a fact that if the grower does everything correctly and at the correct time he will have no difficulty in coping with any conditions that Nature may throw at him.

Weather conditions at the stated planting times can, in a climate such as is met in the British Isles, mean anything; rain, drought, gales or even a late frost. Rain will do no harm, and if one is fortunate enough to have a warm humid atmosphere to go with it, the future success of the fuchsia hedge is assured. Drought or excessively dry conditions will mean that the newly planted subjects will have to be watered. In this case there should be no half-measures and the ground should be thoroughly soaked as well as the plant being sprayed overhead. Gales should have no effect on the still quite small plants if notice has been paid to the paragraphs already written on windbreaks. The late frosts do unfortunately present a problem, but with the modern materials available any gardener is well able to cope successfully with this hazard. Cloches, if available, are ideal to slip over the fuchsias; or, should they be in short supply, or in use in another part of the garden, a length of polythene sheeting laid over the complete hedge will check the severity of the frost and protect the subjects underneath. It must be remembered that in early June the frosts, if any, are of extremely short duration owing to the shortness of the night itself, and the idea behind protection of the fuchsias is the prevention of the possibility of a check in growth. The possibility of any very late frost killing a recognized variety of hardy fuchsia is extremely remote, assuming of course that all the plants were healthy and well hardened off at planting time.

The first summer of a fuchsia hedge is perhaps the most

important part of its life. The gardener has the responsibility of growing the hedge on with as little check to growth as possible, and into a unit which, when the trial of the first winter has to be met, is able to meet the worst weather with a constitution well able to take it. This summer care is what many gardeners regard as routine. Any signs of dryness will mean the use of the hose or watering can, for the fuchsia is a moisture-loving plant. Hot weather may well mean that the leaves will wilt a little despite plenty of moisture at the roots. When this happens it is beneficial to spray the plants overhead as soon as the heat of the day has passed. By five o'clock in the afternoon the sun may still be shining on the hedge, but its power is on the wane and it is quite safe to spray the fuchsias. The sun has only to be feared in the morning, when its power is increasing rapidly, and at noon and in the early afternoon when it is at full strength.

Pests must not be allowed to weaken the hedge by sucking the energy from the plants, therefore a regular spraying programme with insecticides is essential. The mistake made by so many is to spray as soon as the pests are seen. Most pests are first noticed by the damage they have already done to the subject being grown rather than being observed themselves. Pest control is one of the few gardening operations which is carried out by the clock rather than being controlled by the conditions prevailing.

A newly planted hedge should need no fertilizer during its first year of growth, but in its subsequent years it will benefit greatly from heavy mulching and regular feeding. By the autumn of the first year a hedge of quite reasonable proportions should have developed and with the coming of winter it is well able to stand its first real test. A fuchsia hedge that comes through its first winter without casualty can be regarded as established, and will be practically trouble-free for many years to come. However, the first winter is its real test of hardiness and the ultimate outcome of this test may well reflect the care the gardener has given during the initial summer and autumn.

There are several ways in which the grower can help the young hedge to survive the winter successfully and all at little cost. The easiest method of protection is by earthing the plants up 6 inches up the main stem. This is a crude method but it does prevent damage to what at this stage would be a very shallow root run, and partly protects the leaf buds at the base of the shrubs. An improvement of this method is to use sifted ashes for the same purpose instead of soil, and an even greater refinement is the use of peat which, if applied in a very dry state, does take a considerable time, sometimes the whole winter, to become wet right through; and this dryness means that the roots and the base of the plants are insulated from the cold.

Perhaps the best method of winter protection is to use bracken. The bracken is not only laid over the ground for root protection, but interwoven roughly throughout the superstructure of the hedge to afford some sort of frost protection throughout its length and breadth. This is a job which can be quickly done, and the bracken network can be quickly dismantled and dug in when the growing season starts once again. There should be no hurry to remove the winter protection in the spring. Late April or early May is quite soon enough, depending on the weather. If the weather is very bad no harm will be done if the protective materials are left on until later in May.

Once growth has started it is natural for the grower to think of pruning the hedge. This urge for tidiness should be resisted, as the time to prune the hardy fuchsias growing in the open is in the middle of June. Some branches of the hedges will be lost in the winter frosts, and in some parts of the country the fuchsias may well be cut down to ground level. Pruning should be left until the grower has ascertained exactly from where on the plant the new growth is coming; and the old growth, if it has succumbed to the frost, should be left on the shrub for as long as possible to afford a little protection to the new growth.

It is with the coming of the second summer that the grower will fully realize the wealth and colour of his hedge. Those growers whose hedges have come through the winter practically unscathed will find, depending on the variety, that they will now have a hedge of some considerable size; whereas those whose fuchsias were cut down to ground level will have a lower hedge, but will be surprised at the density and the speed of growth of the shoots that come up from the root stock.

It is now that the fuchsias will derive full benefit from a feeding programme. The fuchsia is a gross feeder and because of this its wants should be regularly attended to. A good general fertilizer, of which there are dozens on the market, will satisfy the needs of the plants; and provided that the manufacturers' instructions are strictly adhered to good results can be expected. A great deal of fertilizer is wasted through lack of humus in the soil, and the provision of a good compost mulch in the spring will not only help to overcome this but will help to provide the cool root run so essential to healthy fuchsias.

Pruning is an operation which will be guided more by circumstances than by knowledge. Growing conditions may well limit the ultimate height of the hedge, but where the winter has left the hedge unscathed it would be well to cut out some of the older wood in order to encourage the younger growths. In all cases pruning should mean the removal of all dead wood.

CHAPTER III

Fuchsias for the House
and Window Sill

The boom in house plants died with the First World War and was reborn after the Second. The new boom did nothing for the fortunes of the fuchsia. In the years before 1914 one of the most popular pot plants for both house and conservatory was the fuchsia. The present boom, while favouring a number of choice and most suitable plants for home decoration, is also being exploited by the introduction of species of plants which are very unsuitable for this purpose; in fact, far less suitable than the fuchsia, which is being neglected.

The people of the Victorian and Edwardian era seemed to understand the needs of flowering plants introduced into the house; nowadays, many are under the false impression that indoor pot plants should consist of leaves only, some decorative, some plain and austere. Consequently, the vast majority of plants being sold for house decoration at the present time have no more, and some have even less, decorative value than the aspidistra which most people, for reasons unexplained, despise.

The art of growing flowering subjects in the house is the art of understanding the subjects being grown. Our ancestors grew successfully indoors fuchsias, zonal pelargoniums, cyclamen, and a host of other flowers, even in the presence of that arch-enemy of all plant life, coal gas, but their understanding was such that their plants lived and flowered in spite of this draw-back, to the pride and joy of their owners.

To grow the fuchsia indoors requires a special technique, as the only natural condition it has in such surroundings is shade,

41

and even that may be missing if the fuchsia is placed by a large window facing south.

The first thought in indoor growing should be choice of variety. The variety is important if only for the fact that some varieties will retain their flowers and buds even in the driest conditions, while others will immediately drop every flower and bud at the first suggestion that they should move indoors. Some people not wishing to bother about sending to a nursery for one or two plants may make a purchase from a local market stall or chain store. It is worth noting that the vast majority of fuchsias offered for sale in this way are the varieties kept alive from that era when every other house had its fuchsia plant. Varieties such as 'Fascination', 'Ballet Girl', 'Mrs. Marshall', 'Scarcity', and 'Achievement' have lived as room plants since Queen Victoria ruled the Great Empire, and they are among those still offered for sale in local shops. They have proved that they can hold on tenaciously to life and are therefore quite a sound buy.

Having purchased the plants either locally or from a specialist nurseryman, the next requirements are the pots and the compost in which the fuchsias are to grow. The choice of soil or compost presents no problems as this can be purchased from any horticultural sundriesman or local store. With the sundriesman one must specify that J.I. Compost No. 2 is required, but the polythene bags from the local store merely labelled 'Potting Compost' practically always contain the No. 2 Compost and will therefore do the job required.

The choice of pot is perhaps a little more difficult especially for the housewife who will always have her eye on the most decorative even if it is not the most practicable. For the indoor fuchsia the traditional clay pot can be used either on its own or in conjunction with plastic pots. On the other hand plastic pots can be used on their own. The clay pot is too well known to warrant description, and for proof of the success of growing fuchsias indoors in this type of container, one has only to look back to the beginning of the twentieth century when the fuchsia was enjoying its heyday.

Plastic pots are usually of three types, the hard plastic, the soft plastic, and the expanded polystyrene lightweight type. As these containers are of such recent introduction it may be well to look into whatever advantages or disadvantages they may have, or are claimed for them, as against the traditional clay pots. Taking first the hard plastic, depending on the manufacture, some are inclined to be brittle, and when handling a plant in one of these pots it is essential that two hands be used. To pick up the plants by the pot's rim is courting disaster as the weight of both compost and plant will invariably prove too much for the pot. Furthermore a great number of pots of this type are offered for sale in the most garish colours and it is well to remember that only in the most neutral-coloured containers will the full beauty of the fuchsias, or any other flower, be seen. To consider their advantages, it would appear that the greatest one is their power of conserving water. Fuchsias by nature of their growth, and under some indoor conditions, are apt to dry out quickly when grown in clay pots. This is partly due to the loss of moisture brought about by the porosity of the pots. Also this passage of moisture through the sides of the pots is believed to cause the well-known habit of roots immediately striking from the plants to the sides of the pots, and from there making their way round and round the sides, leaving a large area of almost unused soil in the centre of the pots. Another reason for this is undoubtedly the difference in temperature between the mass of moist soil in the centre and the warmer air passing through the pot from the outside. Plastic pots are warmer in themselves, and being non-porous maintain a much more even temperature throughout the compost. The main advantages are therefore that less water has to be given to the plant and that the plant makes a far greater use of the volume of compost in which it is growing.

The softer type of plastic pot although unbreakable is (because of the softer nature of the material) very difficult to keep clean. Dirt gets ingrained in the plastic and the pot becomes unsightly. Here again manufacturers have given us colours most unsuitable for the growing of flowers.

The expanded polystyrene flowerpots are a little more expensive than other types of plastic pots, but are proving in many cases the most suitable for household use. Once again it is essential that both hands be used when handling these pots as their only fault appears to be a lack of strength. As a finished article they consist of approximately 10 per cent material and

90 per cent air. As static air is only second in insulation value to the vacuum, it can be seen that this pot is the warmest of all, with hardly any variation in temperature, whatever the conditions. This type of container is eminently suitable for use in conjunction with the clay pot. The polystyrene pot used as an outer container for the clay pot will greatly enhance its appearance and certainly assist the clay pot in maintaining an even temperature within itself.

For those who prefer to see their fuchsias in the traditional clay pots—and tradition dies hard in the gardening world—there are two ways in which their moisture-retentive properties can be improved. First, when potting into the flowering-size pot, the clay pot can be lined on the inside with the thinnest polythene sheeting, 150 gauge; and second, they can be glazed on the inside with a material called 'polyurethane'. Most paint manufacturers market this material under such names as 'Polyurethane Glaze' or 'Polyurethane Clear Varnish', and it can be purchased through any high-class decorators' merchant. This material should be brushed on to the inside of the pot in order to seal it completely, and it has no ill effect on plant life whatsoever once it has been allowed to dry thoroughly.

While the choice of pot and compost is of the utmost importance to the fuchsia indoors, perhaps the operation which means most in getting the best from the plant is its acclimatization. It should be easy to see that a fuchsia grown under ideal nursery conditions is not going to take too kindly to the unnatural conditions of a living-room. The change-over process must be gradual, and the failure to understand it has probably resulted in the loss of more fuchsias than any other cause.

The fuchsias should be purchased in the spring and then the plants will have a whole season's growth in front of them. Plants should be young but growing strongly. It is unfair to expect a small, half-rooted cutting to grow into a plant of any size under the conditions it is likely to meet in a room, just as it is unfair to expect an old plant which has lived so long in a greenhouse to survive under exactly opposite conditions.

It is better if the newly purchased plants are already potted into their first-size pot, as any root disturbance is likely to cause a check in growth and a lengthening of time before the flowers come. After all, it is for flower decoration that the plants have been purchased. Some nurserymen and practically all local stores sell their plants with just two or three blooms on the plant. This is an unfortunate practice, but so necessary with a public who will not buy unless they can see what they are buying, as the plants at this stage are normally far too young to stand the additional strain of bearing flowers which ultimately lead to seed and further strain on the plant. At this period of growth, then, all buds, flowers and seed-pods must be removed.

Acclimatization should begin immediately on receipt of the plants. The fuchsias are young and strong, and already in their first-size pot. Their first situation should be the lightest, airiest inside window sill you can offer. A table by the casement or French windows is ideal. If there are already net curtains at the window so much the better, as on dull days the nets can be behind the fuchsias, so that they get all possible light; and should the spring sun become too powerful through the glass the nets can be draped in front to protect the plants. Alternatively if the glass magnifies the sun's heat too much the plants will derive great benefit from the window being thrown open, assuming, of course, that the wind is not too cold. The best room for this purpose is one which is not heated regularly by open, gas or electric fire. Central heating can be beneficial provided that the fuchsias are not too close to the radiators. Spring nights can be very cold and the plants should be drawn deeper into the room, away from the windows, before sunset.

As root growth develops the plants can be potted on into their flowering-size pots. Assuming that the fuchsias when purchased are in size 60, or 3-inch pots, once the roots are moving freely around the outside of the soil it would be quite in order to pot on the plants into the size 48, or 5-inch pots. The only danger the fuchsias run at this stage is from the over-zealous use of the watering can by the grower. When the potting on is done the plants should be given a good watering and then left until almost dry. Roots do not grow freely in a very wet soil but grow prolifically in soil that is only slightly moist. While the plants are developing thus it would be to their benefit if the leaves were sprayed daily with a fine spray. On the market are several types of small hand pressure sprays which are ideal for this purpose. Adjusted correctly and used at the correct pressures they can be used in the house most effectively without damage to furnishings.

For indoor decoration a fuchsia has no need to be potted beyond the 5-inch pot stage. This size pot plus good cultivation will produce a plant of considerable size while still keeping it under some kind of control.

Training a pot-grown fuchsia is always an important undertaking if the grower is to get the best from the plant. For house decoration usually only one pinch, or stopping, is sufficient. If this removal of the leading growing tip is carried out when the plant is 4 to 5 inches high it can be reckoned that at least six breaks will occur, giving six branches bearing flowers. This may of course vary from variety to variety, but if any of the tried varieties listed in Chapter XII are grown this effect will be easy to realize.

Many who have seen the greenhouse and outdoor-grown varieties with their masses of branches and flowers will think that six branches is a meagre allowance for a fuchsia. What must be remembered is that under such artificial conditions as are met in the living-rooms of a modern house, the plants should not be overburdened by your efforts to produce the effect of a fuchsia grown under more genial conditions. Furthermore, a light, airy impression must be aimed for and not the heavy effect of the exhibition plant, which would look ungainly and overpowering among modern furnishings. Indeed, a further argument would be the ease of training and management of the more thinly branched plant. A fuchsia with six branches would need only a thin cane, up the length of which would be trained one branch with the purpose of disguising its presence. From this centre support thin nylon cord can hold the other five branches in position to prevent breakages when moving the plant to dust the room or do the watering. The plant is now supported invisibly, and when the bud and flowers come, the weight will cause the branches to arch in a delicate, pleasing manner.

As already stated, overwatering of the fuchsias will do more damage than good. It is indeed wrong to water any plant on a given time basis, as those plants standing on a window sill in full sun will probably want watering every day, while those in a cooler, shadier position may go for many days without the

need for more water. Practically every gardening book gives
the test for watering, which is to tap the pot with the knuckle,
and if the result is a ringing tone then water should be given.
This operation, however, only refers to clay pots, whose use
indoors is becoming every year less and less popular. It has
already been mentioned that fuchsias grown in plastic pots will
require less watering than those grown in the traditional clay
pots, but watering will still have to be done. The only guide in
this case is the soil and the plant itself. The soil or compost will
itself feel dry to the touch, although this may of course mean
that only the top surface is dry, so that the second test should be
to feel the leaves of the fuchsia. The plant should have stiff,
crisp leaves of a succulent nature and any laxity in them should
immediately be taken as a sign of drought. Very often when the
compost in plastic pots tends to become dry the compost itself
shrinks away from the side of the pot and this also is a sign that
water is needed.

The question of feeding the indoor-grown fuchsias is another
problem that comes up whenever they are spoken about. How
often? How much? and With what shall I feed? In days gone
by the gardening fraternity took a great delight in making the
small task of feeding their plants as great a mystery as possible.
Even in these days of enlightenment there are still many
gardeners who claim to have secret feeds of their own concoction
capable of producing wonder plants. Evidence suggests that a
bottle of general fertilizer bought from the local sundriesman
will often produce as good, if not better, plants as those fed
according to a secret formula. It is the skill in applying these
feeds and fertilizers that produces the real results. That bottle
of fertilizer bought locally has behind it the enormous resources
of great scientific establishments with some of the greatest
scientific brains tackling all the problems that beset the everyday
gardener.

If a liquid fertilizer is chosen for feeding the fuchsias, the
method and measurements of mixing it as described on the
bottle should be strictly adhered to. Should the label list mix-
tures of different strength for various purposes it is always safer
to adopt the weakest mix for indoor plants, whether fuchsia or
anything else. Again the manufacturer can only generalize in
his instructions as to how often the plants should be fed. With

fuchsias grown in the home, feeding should be dictated by the circumstances and growing conditions and not by the day or date of the month, irrespective of what is laid down by the instruction on the bottle or can.

The most important factor governing the feeding of house-grown fuchsias is the amount of watering involved in their general maintenance. The initial compost used when the plants are potted into their flowering-size pot contains a base fertilizer which is normally capable of supporting the plant until the roots are strongly established and the flower buds are ready to come. This is particularly true when the fuchsia is grown in a plastic pot which may well mean that watering will only be necessary once or twice a week. Such circumstances would mean that a weak feed given every three or four weeks would be ample to maintain optimum growth. On the other hand a plant grown in a clay pot situated in a sunny window might need watering once or even twice a day, which means that the base fertilizer would soon be leached from the compost and feeding would be necessary more often, say every week or ten days.

For general use in the home the most suitable fertilizer is that which comes in tablet form. This method of feeding house plants has been used most successfully in the U.S.A. for some years past and there are now several brands obtainable from local sundriesmen and gardening shops in this country. Their use is quite simple, the grower merely having to press the tablets into the compost, and numerous tests have proved that they are quite safe in use. The manufacturers' instructions for their use should be strictly followed.

The everyday maintenance of a fuchsia in the house means only a few moments of one's time. In fact no more time than is necessary to dust bric-à-brac that would stand in the place now occupied by one of Nature's beauties. While the room is being cleaned it would be greatly beneficial to the fuchsia to stand in the outside air for a short time, and if the flowers have still to come a good overhead spray will revitalize the plant and clean it of dust. All dead leaves and flowers must immediately be removed from the plants, as they may be injurious to the plant and they look untidy as well; and when it reaches that stage, a fuchsia will cease to be decorative in the house.

Carefully maintained, certain varieties of fuchsia will be a joy in the house for many months of the summer. Our Victorian ancestors grew them in their homes very successfully and in many of the houses of that age the flowering pot plants were the only items of real gaiety among sombre, heavily curtained surroundings.

There is also no reason why in the following summers the same fuchsias should not again flower and be as decorative as in their first summer, if only the grower will give a little thought to winter care. It is essential for every fuchsia, if you are to get the best out of it, to have a period of rest in the winter. The harder and more difficult the growing conditions the more complete this rest period has to be. The fuchsia is a deciduous shrub, and Nature must be allowed to take its course; although, as the plant is being grown in unnatural circumstances, Nature must be assisted by the grower.

From about the middle of September the fuchsia, even if it is still blooming strongly, should be gradually brought to rest. Little by little water should be withheld from the plant. At no time should the compost be allowed to become dust-dry but eventually only enough water should be given to keep the compost very slightly moist. As this operation will mean the falling of leaves and the dropping of flowers and buds, it would be much more practical if the plant were placed somewhere out of doors. Being in the open would also allow the existing wood to ripen and thereby assist its survival through the dark months ahead. It would only be necessary to bring the plants indoors when frost threatened.

When the fuchsia is almost denuded of buds and leaves it must be brought indoors for its winter protection. It is admitted that a few bare twigs, which is all that the plant will now display, will not be a decorative feature in the house or flat, so assuming that no greenhouse is available, an out-of-the-way corner must be found for it. A frost-proof garage, shed or coal cellar is ideal if the fuchsia is not just put in there and forgotten. The plant should be looked at once a fortnight in order to assure oneself that the compost has not completely dried out, or that the plant is not breaking into premature growth. This will mean that the winter care of the plant takes approximately two minutes out of one's time every fortnight, which shows that it is not a very demanding task.

Should the compost be found to be very dry it should be thoroughly moistened. The best method of watering a plant that is so dry is to place the whole plant in a bucket of water so that the compost is soaked. Such a watering would last the plant several weeks under normal conditions. If the plant is found to be breaking into premature growth it will mean that the storage conditions are far too warm, and it would be better for the health of the fuchsia if it could be moved to a cooler place. A great deal of good, and no harm whatsoever, would befall the plant if on the milder winter day, it were stood on the outside window sill, the veranda, or some handy place in the open.

The grower should not be too anxious to start the plant into growth in the spring. Late April or early May is quite soon enough, as much better results are achieved if the plant is allowed to grow strongly from the start instead of allowing it to become drawn and spindly through trying to grow it during some of the very dull weather experienced in early spring.

The fuchsia, brought out of its winter storage, is once more placed in a light airy position. As soon as growth starts the plant should be re-potted into a fresh clean pot. To obtain the best from the new growth, as much as possible of the old soil and dead root from the previous year's growth should be removed and replaced by fresh compost. During this operation many of the newer roots, which can be recognized by their whitish appearance, will be accidentally broken off and this may cause alarm to those not used to the fuchsia. The more experienced grower will realize that their loss is of no serious consequence, perhaps slowing down a little recovery from the shock of re-potting, but nothing more serious than that.

When the check of re-potting is overcome, and growth is again under way, it will be necessary to prune the plant. For the sake of its future appearance it is better to prune the branches down to the lowest active growth, as nothing is more unsightly than a few growths appearing from the top of a whole length of unproductive wood.

The yearly cycle is now complete, with the little routine tasks carried out last year being repeated this year; and there is no reason why the fuchsia or fuchsias should not continue to flourish.

Fuchsias for the House and Window Sill

It is encouraging to all those who have a natural love of flowers to see so many architects and authorities insisting on window boxes being introduced to what would otherwise, in many cases, be uninspiring concrete blocks of flats. People in influential positions are beginning to realize that flowers, shrubs and trees play an important part in modern town development. While it is true that many of our larger towns just have no room for houses with gardens, it would have been a disaster if the planners had failed the public by not providing colour in the open spaces, and the natural colour that only flowers can provide for the home itself.

Unfortunately the public themselves have still to be educated in the all-the-year-round use of window boxes, and as the authorities have not the labour, money or time to maintain the boxes themselves, the effect is so far disappointing. However, the utensils are there and it remains for the Gardening Clubs and Societies to undertake the task of educating flat dwellers. Such a task was successfully tackled many years ago, when vast numbers of slum dwellers who had hardly ever seen a garden were moved into houses with gardens on huge estates. Many of those same council house gardens are, for their size, some of the most decorative and productive gardens in the country. As one section of the public was educated in the use of flowers and plants some years ago, so the town's new flat dwellers can be educated in maintaining their window boxes, not only to their own benefit, but for the benefit of everybody.

The fuchsia is an ideal plant for use in conjunction with window boxes, or for growing in pots on the window sill or balcony. Here again the grower has the benefit of many months of continual flowering with the minimum maintenance. The little actual time spent in growing a window box of fuchsias would in no way interfere with viewing television or outings in the car. The amount of time any one person would have to spend daily to maintain, say, two or three window boxes devoted entirely to the fuchsia, would be at the maximum five minutes.

Here again some guidance must be given as to the varieties to grow. Tall, upright-growing varieties might tend to grow too strongly and obscure a little of the light to the windows. The larger-flowered types would have their lovely blooms bruised and marred, or even blown off because of the exposed position of the window boxes themselves. Again, some varieties have a rather brittle structure which would mean broken branches and a disappointing display. The varieties listed in Chapter XII are well-tried ones which are sure to give a first-class display if the elementary rules of culture are followed. As these varieties are not generally those sold in the local markets and stores, it would be best to start the whole course of operations by an order to a specialist fuchsia nurseryman. In this way the grower can be certain that what is received is exactly that which is ordered. The order should be in the nurseryman's hands early in the New Year with details as to the purpose to which the plants are to be put. If it has been explained that the fuchsias are to be grown in a window box, and that the grower has no other means of growing the plants other than in this box, the nurseryman will ensure that only well-hardened plants are dispatched to his customer, at a time safe for him to plant straight into the growing position, which is usually the end of May or very early in June.

Before placing the order the grower should give thought to the effect he wishes to obtain from the window box, balcony or wherever he is growing the fuchsias. What he should aim for is getting as much flower and colour as possible from the limited space available. To take as example a window box 48 inches long by 14 inches wide: first let it be said that for ease of management, and planned effect, the whole box should be given over to one subject, in this case the fuchsia. Along the edge nearest the window the dwarfish but very bushy fuchsias should be planted. Allowing 12 inches to each plant this could be planted with two 'Alice Hoffman', with their red sepals and white corollas, used alternately with two 'Mr. A. Huggett', which have cerise tube and sepals and pinkish-mauve corollas. In front of these fuchsias should be planted trailing varieties which will cascade over the sides, giving a band of well-defined colour and disguising the harsh outline of the window box itself. Merely as a suggestion let it be imagined that here again four fuchsias are planted, two 'Marinka', used alternately with two 'Golden Marinka'. This imaginary window box is now planted with eight fuchsias all of which will, under practically all conditions, bloom profusely throughout the summer, Many owners of window boxes, perhaps more artistic than the author, could think of many more schemes to try, but there are few, if any, other genera of plants with the adaptability and the form to give as pleasing a display as the fuchsia.

The essentials for growing fuchsias in this way are the same as for growing fuchsias anywhere, i.e. good drainage coupled with a compost of good moisture-retentive properties. This will mean that the bottom inch or two of the box will be taken up with drainage material such as broken crocks, and this is covered in turn by pieces of broken turf which will prevent the compost in the upper layers running into the drainage material and upsetting its drainage properties. These pieces of well-rotted turf are easily obtainable from horticultural sundriesmen, as broken down they form the bulk of the J.I. Composts which they make up for sale to the public. At the same time the grower should purchase the J.I. Compost No. 2 which will be the growing medium for the fuchsia. Although the J.I. Compost No. 2 should not be rammed down, it should be packed firmly. The window boxes will, at some time during the summer, be subjected to gale-force winds, and loosely packed soil might well mean that complete plants are blown away. Furthermore the watering that will have to be done will cause the compost to settle after the fuchsias have been planted, leaving the plants' roots exposed to the elements and causing their subsequent'loss.

The fuchsias should be planted firmly and given cane or stick support immediately. Do not give each plant a cane just large enough for its immediate use, as throughout the summer, as growth continues, the leading growth should be tied up to prevent breakages through storms and gales. The trailing varieties which are to hang over the edges of the window boxes should be given a frame over which they can scramble (see illustration) and this will not only help to prevent loss through breaking but will also prevent the flowers being bruised by the wind blowing the blooms against the sides of the window box.

Watering presents no real problem to the window-box grower, as one has only to scrape a little of the top soil away to see if the underneath soil is moist or not. What the grower has to be really careful about is that at no time are the fuchsias allowed

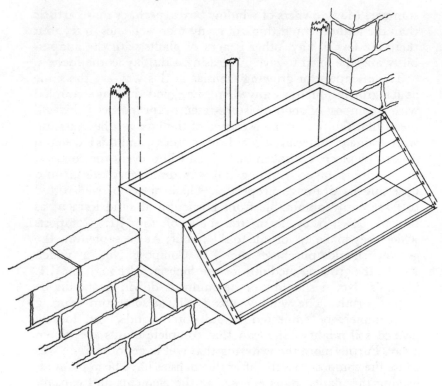

Window Box With Frame To Support Trailing Fuchsias.

to get dry. This applies to all aspects of fuchsia growing, but with window boxes there is an added danger from the weather. Once a plant is allowed to become limp it has no natural protection from whatever wind may be blowing at that time. It is unfortunate that window boxes tend to dry out more rapidly when a strong wind is blowing, but if the grower understands this he will be forearmed against these dangers.

This is perhaps the time when a plea must be made to flat-dwellers for courtesy. Architect-planned flats have window boxes that are normally planned so that excess water from the box in no way interferes with the comfort of people living below. Home-made window boxes are not always so well designed, and it is common courtesy to channel the water so that it does not annoy passers-by or those living beneath the box of flowers.

A little thought should be given to the feeding of the fuchsias. Over-generous feeding may lead to the plants' making heavy, sappy wood and large leaves. Owing to the conditions to which the shrubs may be subjected this can be a great disadvantage, so it may be better to underfeed rather than overfeed the plants. A weak feed once a fortnight will prove ample and will sustain the fuchsias throughout their growing season.

Training of the bush-type plants should aim at producing a round bushy plant with short growths rather than allowing the normal fuchsia habit of long arching stems to develop. This can be achieved by pinching out the leading growths once or twice more than is usual when growing bush fuchsias. The pinching should stop by the first week in July, or many weeks of flower may be lost. The trailing fuchsias, if they are chosen to adorn the window box, should not be stopped so often but should be allowed to follow their normal habit, which is to hang downwards.

Growth would greatly benefit if the plants were sprayed overhead daily. One of the small hand pressure pumps recommended for use in the house would again be ideal for this purpose. Do not spray in bright sunlight, but wait for the shadow to fall over the window, and then give a good spraying to both the top and underside of the leaves. This regular spraying will also deter attack by insect pests. If, however, a pest attack should come, no time should be lost in dealing with it. A window box is continually in full view of everybody, and because of this neglect will immediately show itself. For ridding the plants of pests there is no handier weapon than the aerosol garden spray. Several firms now have these canister-type sprays on the market and they can be purchased from the local stores or florist's. They are ready for instant use and are deadly in their effect on all insects that are likely to attack fuchsias grown in a window box.

With the coming of autumn the owners will wisely start to think of planting their bulbs, and that will mean the removal of the fuchsias. They must be taken in before the frost strikes, and put into pots for the winter. It may be that a grower who lives in a flat will not have, or even want to have, empty pots lying around all summer just to put the fuchsias in in the winter. In this case bitumen paper pots could be purchased quite cheaply and they would contain the fuchsias and keep them safe until they take their place in the window box again at the latter end of the next spring.

Winter care is the same as that recommended for fuchsias grown in the house.

Growth should be started again in the early spring, about the middle of April. The plants should be pruned to a neat shape, and for convenience they can be stood during the day in their pots among the bulbs in the window box. They will in no way interfere with the display of the bulbs but will themselves benefit from the milder spring weather while their spring naked-ness is concealed. At night they should be brought indoors and given the protection of the room, since a frost, while it may not kill the fuchsias, may well do enough damage to ruin the whole year's display. By late May or early June, depending on the locality, the growing plants should once again be placed in their positions in the window box.

To get the finest results from the fuchsias again it is essential to have fresh soil in the boxes every year. It must be remembered that for a whole twelve months this relatively small amount of soil has had to sustain growing plants, first fuchsias and then bulbs. Such extensive cultivation must quickly exhaust the original goodness of the compost, and no matter how much fertilizer one may put in to replace that which has been used by the plants and bulbs, this will in no way return the compost to its original, well-balanced structure.

What has been attempted in this chapter is to bring to the smallest grower the need for an understanding of the simple requirements of the fuchsia when it is introduced into the home. 'Green fingers', the gift which everybody's neighbour has, is merely the application of that understanding and knowledge.

CHAPTER IV

Fuchsias for the Open Garden

During the eighteenth century tremendous stress was laid on the geographical locality of plant life, although explorers had not yet been able to give details of the actual environment in which the plants grew. It was assumed by almost everybody that everything on the Equator flourished only as a result of exposure to great heat; and it was known and believed only by the most advanced brains that at some parts of the Equator there was snow all the year round. It was this misunderstanding which led to the hardiest of fuchsias, *F. magellanica*, being introduced originally into this country as a greenhouse shrub.

While it was eventually found that some species of fuchsia were hardy enough to withstand the English weather throughout the year, practically all gardeners were afraid to risk the more flamboyant hybrids in the open garden except during a few weeks in the summer when they were bedded into specially prepared flower beds, or stood in their pots on the porches or terraces of the grand houses of that day and age. During the First World War, when all forms of artificial heat were lost to plant life that did not assist the war effort, a handful of the keenest gardeners discovered many of these so-called greenhouse varieties were quite capable of withstanding the notorious English winter. Unfortunately the publicity needed for conveying to the post-war gardeners the information gained was not forthcoming, and the fuchsia, from a gardening angle, went into decline.

At the outbreak of the Second World War the fuchsia had a small but very keen following in the newly formed British Fuchsia Society, and once again the fuchsia had to stand the rigours of the English climate without the aid of any artificial heating. This time, however, the B.F.S., under the forceful leadership of its secretary Mr. W. W. Whiteman, was quick to record and proclaim the hardiness of their favourite flower. It was found and recorded by the Society that a vast number of the thousands of varieties of fuchsia would survive the milder form of winter unprotected in any way, and given some form of simple covering would survive and flower after the most severe conditions.

These findings led to experiments being carried out simultaneously in Oregon, U.S.A. and in this country, on the growing of fuchsias without the aid of either greenhouse or frame. Unfortunately the experiments in this country had only just got under way when Mr. W. W. Whiteman died. His work was quickly taken up by Mr. W. P. Wood, who proved conclusively, by growing hundreds of varieties without the aid of glass structures, that not only will fuchsias grow without artificial heat, but that they can be thus grown up to exhibition standard. Some of the finest displays of fuchsias ever exhibited by an amateur at the Royal Horticultural Society's Halls at Westminster were those put up by Mr. Wood from fuchsias grown in the open garden. Although his sudden death occurred while experiments were still going on, much of his work was recorded, and many fuchsia enthusiasts are using his methods to grow the flower of their choice without the expense of greenhouse and frame.

The chapter on 'Hedges' gives details of the culture of the recognized true hardy species and varieties. This chapter is mainly concerned with the growing of the so-called 'half hardy' varieties of fuchsia outdoors. The greater number of fuchsia hybrids are suitable for this form of culture, the most notable exception being the group popularly known as the 'Triphylla Hybrids', of which perhaps the most common bedding fuchsia of all, 'Thalia', is a member.

There are two methods of growing the shrubs in the open. One method is by keeping the plants in flowerpots, and the second method is by planting the subjects in permanent beds. I do not propose to give the cultural details of the fuchsias to be planted into permanent beds, as this would merely be a repeat of the culture of the hardy fuchsia, which is described in Chapter

Fuchsias for the Open Garden

II. Again, since I intend to give the complete cultural procedure for pot-grown fuchsias in the next chapter, which deals with 'Fuchsias for the Greenhouse', I shall confine myself to describing the cultivation of plants grown under the described conditions and dwell especially on their care under winter conditions.

Even the most tender species of fuchsia in its natural habitat is a shrub of the cool regions. Geographically a large number of these species grow within the torrid zone, but it is a fact that the nearer a fuchsia grows to the Equator the greater the altitude at which it is to be found. Those species whose natural geographical location is the Equator itself would be found at 7,000 feet or over. At such altitudes the plants are subjected at times to quite severe cold, and a great deal of moisture from low-lying clouds. It is believed by many very experienced gardeners that the greatest enemy of the fuchsia is drought, and the genus is capable of withstanding any degree of cold provided that it has facilities for taking in moisture. It is claimed, and not without foundation, that when sub-zero temperatures are able to freeze the moisture round the roots of a plant, the plant itself suffers from the fact that it is unable to take in the now solid moisture, and a condition of drought ensues. The prolonged severe weather of the winter of 1962–3 took great toll of shrubs on thin shallow soils, whose cellular structure is known to be unaffected by extreme cold. Established bushes of such shrubs as buddleia and cotoneaster succumbed, it is said, more through the fact that there was no available moisture for many weeks than through the effect of the frost on the cells of the plant.

The methods of overwintering fuchsias carried out by W. P. Wood in England, and by the American growers in the State of Oregon, have since been adopted by many fuchsia growers, and the reports of all prove conclusively that there is no safer way of overwintering fuchsias than the outdoor method.

According to weather conditions, but usually about the second week in September, a trench is dug in the garden in the position most sheltered from the coldest of the winter winds. The length of the trench should be such as to take the plants comfortably standing in a row with the top branches just touching, and the width should give a 2-inch clearance either side for the widest plant. The depth should be enough to cover completely the tallest plant except for the very tip of its topmost branches. The next step to consider is drainage, and this is best achieved by putting along the bottom of the trench a 1-inch layer of twigs, broken sticks, straw or any rough material which will keep the plants, still in their pots, away from the floor of the trench. On heavy soils where there is no danger of the sides collapsing through heavy winter rains there will be no need to line the sides of the trench; but on light and sandy soils it is often safer to line the sides only with polythene sheeting. The polythene sheeting need only be a 250-gauge material, weighted by soil or stones at the top of the trench and allowed to hang down the sides. On no account must an impervious lining be given to the floor of the trench as this will prevent the rain and melting snow running away normally, the fuchsias will become waterlogged, and losses will occur.

At the first indication of frost the plants should be well watered and placed in the trench and covered with leaves, peat, or any material that will assist in insulating the trench and the plants inside it. Leaves are probably the cheapest and most easily obtainable material for this purpose as the countryman will have no difficulty in collecting them himself; and the town-dweller will often find that a telephone call to the Highways Department of the local council will quickly result in a lorry load being delivered by an authority only too pleased to get rid of them. Whatever material is used, it should be well worked into the framework of the plants as the trench is filled up, although no attempt should be made to consolidate the filling. When the trench is filled to ground level a small mound of earth should be placed above its whole length and a length of polythene anchored to prevent the heavy winter rains from consolidating the filling material too much. It must be remembered that static air is one of the greatest insulating elements known to mankind, and the more the grower is able to trap inside the trench the greater the degree of insulation he will get.

There are, naturally, modifications to this system of overwintering. One successful grower lays his plants in the trenches instead of standing them in it, while another whose garden is favourably sheltered only half-buries the plants and allows the winter frost to prune back the top growths. Whichever way is chosen by the grower it must be stressed that in all cases, to be one hundred per cent effective the top rim of the pot must be at least 5 inches below soil level. Such a depth will protect the root system and the lower growth buds from the severest weather.

At the end of March in the southern part of England or the beginning of April in the north and Scotland, the trench fillings can be removed, as the plants in some cases will already be making growth. This early growth can be very brittle, and as there is no heat to assist growth it should be preserved so that there is as little delay as possible in bringing the fuchsias to maturity. The safest way to remove the filling is to dig away the end of the trench and work along the length, gently shaking the material away from the framework of plants.

At this time of year the danger of frosts is still present; so as soon as the plants have been pruned, the material within the trench should be excavated to such a depth that the newly pruned fuchsias have the tops of their shortened branches just below soil level as they stand on the peat or now rotting leaves. As darkness falls, the polythene, which should now be fixed to a framework of laths, should be anchored over the top of the trench. This covering, together with the heat being generated by the decomposing material underneath, will protect the fuchsias from spring frosts. Such protection should also be given during those days in early spring when the cold, drying east wind is blowing.

As growth appears the fuchsias should be re-potted and re-placed in the trench immediately, but this time they should be stood on a piece of slate or wood to prevent the entry of worms into the pot through the drainage holes at the bottom. It is a good idea to place over the drainage holes inside the pot, and under the crocks, a piece of finely perforated zinc of a type that can be purchased in sheets at the larger ironmongers', and which is normally used to cover ventilators and food containers such as meat safes. This form of protection is very necessary when later the plants are bedded in the garden up to the rims of their pots, when the danger of frost is over.

Whenever the weather is favourable the fuchsias should be given the maximum amount of air and light. The sun at the early part of the season is not so powerful in its effect as to cause harm to the plants, so all forms of protection should be put to one side on bright spring days.

Frosts of any severity should be past by the end of May and it is then that the fuchsias should be removed from the trench and bedded into the garden. If the plants are later required for exhibition purposes or for decorating porches or summerhouses they must for easy handling be left in their pots. The ideal position in the garden from this period onwards is one of coolness and shade. Although the fuchsias themselves should in no circumstances be placed under trees, they do appreciate the cool shade that large trees will throw when the sun is hot, and also the protection they will give from summer gales. If a place for the fuchsias can be found that is protected in such a fashion, the task of summer maintenance will be made that much easier.

Having chosen the position the plants are to occupy throughout the summer it is now necessary to sink the pots into the soil up to their rims. The purpose of this action is threefold. First, the roots are maintained in a cool condition; second, the wind cannot blow the pots over and damage the plants; and third, if clay pots are used there is no loss of moisture through the pots and the labour of watering is therefore considerably reduced. There should be a distance of 30 inches between pots, for well-grown fuchsias treated in this hardy manner can develop into quite large plants.

The grower will not need to tie or support the plants any more than those that are grown in the greenhouse, for although they are exposed to all the varying conditions of an English summer they develop a harder growth which is well able to support itself. Generally all that is necessary is one cane, to which all lengthy growth can be loosely looped with raffia.

Cultivation now follows on the same line as fuchsias grown in the greenhouse. Most growers find that outdoor-grown fuchsias are less subject to attack by pests than their greenhouse counterparts, but a regular spraying programme should still be carried out to prevent even the mildest attack taking place. The feeding programme described under 'Greenhouse

Cultivation' is the one necessary to maintain pot-grown garden fuchsias in tip-top condition.

There is practically no garden to which the described form of hardy cultivation cannot be applied. Large collections of fuchsias can be grown and enjoyed without the expense of glass and heating.

CHAPTER V

Fuchsias in the Greenhouse

When the fuchsia was first introduced into Great Britain it was considered to be a greenhouse shrub. Experience over the years has since proved that many species and varieties are quite hardy under most of the conditions they are likely to encounter in these islands, and indeed many of the recognized hardy fuchsias give only of their best when submitted to the rigours of an English climate. Despite this fact, the possible range of fuchsia growing can only be enjoyed to the full when a greenhouse is available with sufficient heat to permit the plants to grow on during the winter months.

The type of greenhouse and the amount of heat needed will depend on what the grower wants from his plants. In no case should a winter temperature of 55°F. be exceeded, but if climbers, espaliers and some of the more delicate species are to be grown, then 50° is the ideal maximum winter temperature. When giving thought to the heating of a greenhouse consideration should be given to the growing of fuchsias in the form of climbers or any special shape that will take them near to the actual glass. Sudden drops in temperature will quickly draw the heat away from the area of the glass, and although the fuchsia will survive as a plant it may well be damaged in such a way as to need retraining.

Many gardeners feel that the heating of a greenhouse for winter use is a luxury that they can ill afford, but even in such cases, although the range of growing is restricted as compared with the heated house, there still remains tremendous scope; and provided that the minimum temperature does not fall below 34°F. it can be a safe and comfortable home for fuchsias. It does not cost too much to maintain such a temperature just to keep out the frost, for winters of prolonged severity are rare in this country. Naturally fuchsias cannot be expected to grow on at such low temperatures, but they will be safe in their dormancy until the spring sun brings enough warmth to start the plants into growth. Such a period need not mean that the greenhouse is short of flowers, for a good gardener would turn the house over to pot-grown hardy annuals or hardy spring bulbs, which are well able to thrive and flower while the fuchsias rest.

As a greenhouse subject fuchsias are ideal. In a heated greenhouse they can be had in flower as early as April, without the aid of artificial light, which is discussed later, and the flowering can be maintained until November. The cold house can have flowers from May until September. There are indeed few other shrubs that will give such value as this.

The technique of growing fuchsias in the greenhouse holds no mysteries, but many growers and would-be growers become puzzled by conflicting instructions that have been written as to how they should proceed. The reason for this is that all procedures of extreme simplicity can be made more difficult without improving results.

The fuchsia year for the heated greenhouse will begin in mid-February, and for the cold greenhouse a month later, in mid-March. Orders for the plants should be in the hands of specialized nurserymen by the previous late autumn, with instructions as to when the new plants are needed. It is unfortunate that fuchsias are difficult to describe in the amount of space allowed them in most catalogues, and if a well-varied collection is aimed for the grower should try to visit, during the summer, the various places where good collections of fuchsias are on show to the public. Such places as Kew Gardens and Wisley Gardens have good collections and there are a great number of local councils who welcome visitors to see their collections of plants, which more often than not include fuchsias. The shows held by the British Fuchsia Society in many parts of the country are well worth a visit to see the variety and versatility of the shrubs, and if one could, by appointment, visit any of the specialized nurseries during the peak of the season there would be no need to ponder over the abbreviated descriptions in the catalogue.

If the fuchsia has a fault it is that it does not travel well, and this is a further argument in favour of those nurserymen who specialize in this flower, for they are artists in packing and dispatching without causing breakages. The grower will receive the fuchsias in plastic or paper pots and it is essential that they be unpacked with the minimum of delay. There is no way of knowing what conditions the plants have travelled under, so the grower can be forgiven if at this stage the plants are 'mollycoddled'.

The plants on receipt should be put into a close, damp atmosphere away from sunlight to ensure the most rapid recovery. It must be assumed that the fuchsias enjoyed, up to the time of dispatch, the congenial conditions of the nurseryman's greenhouse; and even if it is not possible to copy those conditions exactly, the finest conditions available should be given so that the shock of change is minimized. The actual loss of fuchsias during dispatch from the nurseries is negligible, but many growers fail to get the maximum results from their plants through lack of attention on receipt. So often the nurseryman's stock is blamed for lack of success when the fault lies with the purchaser.

Even at the period of optimum growth the fuchsia does not enjoy being placed directly into strong sunlight, so until the plants have acclimatized themselves to their new conditions, it is essential for them to be shaded at all times. It is not easy for the amateur who is out all day to anticipate the English weather, and before the period when more permanent shading of the greenhouse becomes necessary, there is a danger that the spring sun will be highly magnified by the glass, and have a detrimental effect on the young plants. A sheet of polythene film arranged to form a shelter over the fuchsias will diffuse the strong light and thereby bring benefit to the plants, and should the grower already have his greenhouse lined with polythene to conserve winter heat, it would be as well to leave this in position until the early May sun demands the introduction of the summer shading.

A few days in the greenhouse and the plants will soon shake off their bedraggled look and will begin to look perky again. This is the time when they should be removed from the temporary pots in which they travelled, and potted on into their first-size pot. This pot should be no larger than that in which they were grown by the nurseryman, but it is always beneficial to remove a little of the old soil and replace it in its new pot with a small amount of fresh soil. Sometimes on receipt of the plants a great deal of the compost is already knocked off the roots by the jolting received during dispatch, and in this case it is only necessary to pot on the plants with as little root disturbance as possible. Those plants received with their soil ball complete should have a thin layer of the already sour surface soil removed, together with any compost at the bottom of the soil ball which has not yet been reached by the young roots. Do not disturb any compost which is already supporting root-growth.

Until the introduction of the potting composts formulated by the John Innes Horticultural Institution many specialist gardeners took great pride in concocting their own special brand of compost, which more often than not remained for all time on the secret list. The scientific work first started by the Institution has removed many of these mysteries, and the J.I. Composts are practically standard throughout the horticultural world. Time and experiment have proved that fuchsias grow as well, if not better, in J.I. Compost No. 2 as in any other form of potting compost. For the smaller grower it is often more practical to purchase this compost from a reputable sundriesman, but the larger grower and nurseryman who has the apparatus necessary to sterilize the loam may well wish to mix his own. The process of sterilization is most important. Some of the older school of gardening feel that the loam is robbed of much of its richness by this processing, but we now live in an age of sterilization. Growers have long since learnt that cleanliness and hygiene are as essential when combating plant diseases as are the complex chemicals that science has given to us. Unsterilized loam must contain an unknown quantity of harmful bacteria as well as bacteria that will benefit the plants. It is useless to scrub pots, staging and glass to rid them of unwanted foreign bodies and then reintroduce these with unsterilized soil. The beneficial bacteria are reintroduced to the sterilized loam by the aid of the chemical additions to the compost known as the John Innes Base Fertilizer.

The plants, having now been potted, will begin to grow

quickly, and if plenty of moisture is given overhead the growth will be succulent and strong. In the heated greenhouse great care must be taken to ensure that the plants do not take on a drawn appearance. Too much heat on very dull days will tend to draw the new growth, and although such plants will flower in time, their appearance will be spindly and sickly and lacking the sturdiness that characterizes a well-grown fuchsia.

Potting on into a larger pot will soon become necessary. The fuchsia is a rampant grower and the strong root growth will soon become evident. As soon as the roots can be seen beginning to travel around the outside of the compost ball it is time to pot on into a larger pot. Assuming the plants are growing in large and small 60's—pots of 3- and 3½-inch diameter at the top—it is in order to pot directly into 48's, that is pots of 5-inch diameter at the rim. For those growers who have heat to assist the promotion of rapid growth a further potting on may be necessary later in the season, and in this case the 32, a pot of 6¼-inch diameter, will be the pot that will hold the flowering plant for the first year. In the cold house it would be better in the first year to allow the 48 to become the final pot. It must be remembered that potting the fuchsias on into larger pots will produce larger plants but will delay the time of flowering considerably. No plant should be potted beyond the 6¼-inch pot in its first year of growth.

This operation of potting on is very important and should be done without any root disturbance whatsoever. The larger pot should be lightly crocked, three pieces of crock being ample to prevent the loss of soil through the drainage hole, and the crocks covered by a layer of compost which is gently firmed. The fuchsia should be knocked out of its smaller pot, so that the root and compost ball remains intact, and placed in the centre of the larger pot. The existing root ball should be placed low so that it will be covered by at least a half-inch of fresh compost when the operation is completed. The fuchsia, having been put into its position in the new pot, should have the fresh compost worked between the side of the pot and the root ball. A distance of half an inch should be left between the compost and the rim of the pot to facilitate watering.

To the professional gardener the art, and it is an art, of potting on becomes a routine duty, but there is no doubt that many amateurs are worried as to how much the new compost should be firmed. Loosely potted plants suffer great hardship through drought, through their inability to hold water, and the root system suffers in consequence. Compost that has been rammed down too hard forms an impenetrable barrier to the roots and causes stunted growth and premature flowering, because the plants grow as though they are still in the smaller pot. The amateur who has little potting experience should remember that fuchsias enjoy the compost firmed at finger pressure. Forget potting sticks and such gadgets, and use the 'sense of touch' gift that has been given to most gardeners. The compost should be firmed by pressing it with the finger muscles only. Do not use the weight of the body or the power of biceps to reinforce the power of the fingers. If the gardener has not the finger strength of his animal relations that live their lives in trees, he will find that fingers have enough power to firm the compost to the degree enjoyed by the fuchsias.

After potting, the plants should be given a thorough watering, and be shaded for two or three days to ensure a quick settling down in the new pots. Until the roots have taken a firm hold of the fresh compost, great care must be taken not to overwater the fuchsias. It is safer at this stage to maintain the growth of the plants by continually spraying overhead rather than giving water through the pots. Some growers use a weak solution of leaf fertilizer in their spray to assist the fuchsias at this time.

Within a short time the roots will have taken a firm hold on the new compost and the plants will begin to get root-bound in their pots. This will be evident by the mass-production of flower buds, which is followed within a few weeks by a dazzling display of flowers. Once the fuchsia becomes pot-bound it is practically impossible to overwater it, and because of the plant's insatiable capacity for moisture, it is evident that much feeding will be necessary to maintain healthy growth. Fuchsia growers all have their own pet fertilizer and they all get the results they strive for. It appears that any good general fertilizer used at intervals of ten days, once the roots are well established, will keep the plants healthy and induce them to flower well. Many exhibitors adopt the system of feeding their plants with a fertilizer of a nitrogenous nature such as 'Compure', and as soon as the required size of plant is reached they switch to a fertilizer with a higher potash content such as 'Compure K'. The extra potash seems to lead to a more sturdy growth and enhances the richness of the flower colourings.

Throughout the summer months the greenhouse should be kept shaded. Lattice blinds are of course the ideal method of shading any greenhouse, but they require constant attention to raising and lowering according to weather conditions. More popular are the more permanent shadings such as 'Carsolumbra' and 'Summer Cloud', which are sprayed on to the exterior of the glass and which can be removed by a light scrubbing in the autumn.

Spraying the fuchsias overhead will be of great benefit to the plants. On hot summer days this should be done several times during the day to create the ideal atmosphere and to prevent the temperature rising too sharply. It is realized that spraying several times a day is not possible when one has to go to work away from home, but in this case, when hot weather is indicated the house should be thoroughly damped down before leaving for the day. Spray the plants until they are dripping water, and pour plenty of water on to the staging and the greenhouse floor. This will create the damp atmosphere in which the plants will revel. Once the buds are showing colour the overhead spraying should be discontinued, as many varieties, especially the very choice American introductions, do not take kindly to having water played directly on to them. Some mark rather badly and others ball up and refuse to open their sepals. Spraying should still be practised to create the moist atmosphere necessary, but the spray should be directed at the pots in which the plants are growing rather than at the plants themselves.

Watering during the summer months when the fuchsias are in full flower must be attended to daily. In hot weather some plants may need a further watering as soon as the grower returns from his employment, but the golden rule with strong plants during the height of the summer is to water if there is a doubt as to whether the plant will last the hours the grower is away. As already stated, a strong pot-bound fuchsia is a very difficult subject to overwater as its rate of respiration is extremely high and the mass of roots is very greedy; for this reason no stagnant water will collect in the pots.

The fuchsia flowers continuously until late into the autumn so that as flowers fade and drop fresh buds open to take the place of those that die. In order to get the most from the plants the seed-pods should be removed as the flowers drop. On very large plants this may well prove too big a job to be done daily, but if only a few seed berries are removed during each daily round it will assist the plants considerably and help the continuity of flowering. The dead flowers that have dropped to the staging and the floor should be collected and placed on the garden compost heap. To leave them lying about in the greenhouse will only encourage wood lice, earwigs and other undesirable pests, and may encourage *Botrytis*, which is the only real disease that will attack fuchsias. Furthermore the untidiness will detract from the beauty of the plants themselves.

The feeding of greenhouse-grown fuchsias should be discontinued by the beginning of September so that the current year's growth can be allowed to ripen ready for the winter ahead. About the middle of September, when greenhouse maintenance is undertaken, the plants will benefit from a week or so of standing in the open. Being treated thus the wood will further ripen and diminish even more the risk of loss in the winter.

Every fuchsia grower should take advantage of this maintenance period not only to repaint and repair the greenhouse, but to cleanse and disinfect it thoroughly ready for the winter storage of plants.

The plants in the open should now receive rather less water. Already most of the fuchsias will have slowed down their rate of growth and the flowering will be more intermittent, all signs that the plants want to rest. Fuchsias only give of their best when they have had a rest, and it is very wrong to try to maintain growth throughout the winter on any plant except the late summer- and early autumn-struck cuttings. It is possible with sufficient heat and artificial lighting to keep a fuchsia blooming indefinitely, but the plant will eventually become exhausted and useless for further flowering. The gradual withholding of water will accelerate the process towards dormancy without danger to the plant. With the rapidly shortening days and the higher moisture content in the air during September, watering should only be done when the compost in the pot is almost dry.

Plants should be returned to the safety of the greenhouse before the end of September or before that if frost threatens. Some growers will wish to maintain their fuchsias in flower for a month or so longer, but busy gardeners will want the room in the greenhouse and want their fuchsias at complete rest as soon as possible. This is easily brought about by giving only enough water to keep the compost from becoming dust-dry. The leaves and the remaining flower buds will drop, and when the plants are at the stage when they are merely a collection of nude branches they should have the compost thoroughly soaked again and be put in a convenient place for winter storage.

There has, in the past, been great confusion among growers as to how fuchsias can be safely kept in the greenhouse throughout the winter. Old books written by expert gardeners all advocated that the fuchsia be kept dry during its dormant period, while more modern books state definitely that the compost should be kept just a little damp. The old school of gardeners and authors grew wonderful fuchsias but failed to give the layman full details as to the conditions under which the fuchsias were stored. These books were written in the days of the large estate which had its extensive plant-houses and an ample staff to maintain them. The duty of these gardeners was to keep a display of flowers in season throughout the year. This meant that very often the fuchsias were stored out of the way under the staging in an orchid or similar house where the humid atmosphere was enough to maintain life, and any excess moisture would have caused premature growth. For this reason they did not have to water their fuchsias during the winter, and this fact they recorded.

It is now definitely proved that no fuchsia, no matter how dormant, can survive very long without moisture in some form, which means that the grower during the winter will have to strike the happy medium, with enough water to keep the plants alive, but not enough to start them into growth during the mild winter spells. This may sound difficult on paper, but it merely entails an inspection of the plants approximately every three or four weeks throughout the winter, and any that are found to be dry should be placed in water up to the pot rim until the soil is saturated. Such a soaking will suffice for a further three or four weeks according to the weather.

The temperature throughout the winter should not fall below 34° F. Such a temperature in the smaller greenhouse could well be maintained by the use of a paraffin-oil heater. Although the heater should naturally be kept in tip-top condition so as to avoid fumes and smuts, it is worth noting that dormant fuchsias are completely insensitive to any impurities that a heater may be allowed to give off. Should the temperature fall 4 or 5° below freezing point some of the species and the *Triphylla* hybrids may succumb, but this drop in temperature should not seriously affect the general run of hybrids, provided that the sub-zero conditions are not of prolonged duration. However, it is safer all round if the 34°F. minimum temperature is maintained.

For those growers who can command more heat, say a temperature of 50°F., throughout the winter, it is possible to grow on into fair-sized plants the cuttings taken during the late summer and early autumn. There is no doubt that among pot-grown fuchsias two-year-old plants are ideal both for convenience of size and for ease of training. Beyond this age the fuchsia will develop much hard wood which can become unsightly until the leaves and the flowers are grown enough to hide it. The continuous growing on of the summer- and autumn-struck cuttings is a short cut to reaching the two-year-old phase. Furthermore, when growing such trained plants as pyramids and espaliers the task is made much easier by keeping the plants in continuous growth than by having to stop and store them when the weather gets cold, and then hoping that the new growth shoots will come from where the grower wants them the following spring.

It is essential that already established plants should still have a resting period even in the heated greenhouse, and this can be controlled entirely by the amount of water given to those plants. The easiest way to measure how much water should be given is by watching the behaviour of the plants themselves. In the first place they should be dried off sufficiently in the late autumn to cause the leaves and few remaining flower buds to drop. The idea then is to maintain at every leaf axil a small pink bud. Should this growth bud begin to callous over, then more water is needed to keep the fuchsia alive, but should the bud begin to break prematurely then the grower should refrain from watering until the buds once more become dormant.

When storing a completely dormant fuchsia for the winter, it is better for the plant to be rested on its side, for this will encourage growth buds to break from the base of the plant when once again it is started into growth. It is already well known that if an upright branch of any shrub is lowered to ground level, the growth buds just below the actual bend are stimulated and will often produce new shoots. This principle applies to the fuchsia and the act of laying the plant down acts exactly as if the branches are lowered. It is important that these growths from the base be encouraged, for so often fuchsias are seen with lengths of old, gnarled stems surmounted by the new shoots and flowers. Such ill-shaped plants are poor advertisements for both the genus and the grower.

The conservation of heat within the greenhouse is most important these days, when the cost of producing that heat is so high. A greenhouse built with double-glazed units is the ideal thing but the initial cost can be prohibitive, so most growers use the cheapest and nearest equivalent, which is polythene sheeting. This material can be purchased in several thicknesses, but as the growers only require to form an inner skin with a layer of static air between it and the glass, then nothing thicker than 150 gauge should be used. This thickness will serve the purpose of keeping a layer of air static without the loss of too much daylight.

In the heated greenhouse the resting period can end as soon as the grower feels that the day length is sufficient to give, in conjunction with the heat available, a balanced growth. This is usually about mid or late February. Starting the fuchsias into growth would merely mean giving them sufficient water at the roots and an occasional spray overhead. This operation of overhead spraying must be done with care at this early time in the year. Spraying should be carried out on bright days only and then only in the early morning. The air during some of the dark cold days of February has a very high moisture content and this alone will build up to quite a large amount of condensation within the greenhouse. To spray the fuchsias during a period of high humidity would mean that the same water would rest on various parts of the plants for some days, and any spores of *Botrytis* that were in the air would have an ideal breeding ground.

Starting the growth in the cold house, or house where frost is just excluded, should not be attempted until the end of March or early April. By this time the sun through the glass will have raised the temperature in the greenhouse considerably and the night frosts, because of the lengthening days, will be of shorter duration and of less severity. The plants in such a greenhouse should be thoroughly watered and overhead spraying should be carried out daily. This is now possible because the sun's heat is normally sufficient to remove any excess condensation from the greenhouse, and in any case the harder wood of plants stored throughout the winter in the cooler conditions will be far more resistant to the attack of any *Botrytis* that may be present than those fuchsias stored in the heated greenhouse.

The pruning of the plants should be delayed for a short time until the new growth buds have developed sufficiently to indicate where the new shoots are coming from. Nothing can be more annoying than pruning a fuchsia with a certain shape in mind and finding growth developing from the places that have not been planned for. How the plants are pruned will depend on the type of plant desired, but every effort should be made to make them symmetrical so that their beauty is the same at whatever angle they may be seen. Old wood that will detract from the beauty of the fuchsia should be cut right away, and the growth from the base, in the case of bush plants, should be given every encouragement so that the colour of the flower and leaf is present from the top to the bottom of the plant.

As soon as growth is under way the plants must be re-potted into fresh compost. The difference between re-potting and potting on is fully understood by all experienced gardeners, but to the beginner it often causes some confusion. Briefly, potting on is the operation whereby plants in full growth are moved on to larger pots with no, or hardly any, root disturbance at all. Re-potting is a much larger operation whereby a dormant plant, or one just breaking into growth, has as much as possible of the old, sour compost removed before being put into a clean pot with fresh compost.

It is essential that the grower understands fully the principles involved in re-potting as it applies to fuchsias, as it is almost certainly one of the most important operations in the fuchsia year. Re-potting takes place after the plants have been pruned to shape and the new shoots have made one or two pairs of leaves. At this stage the plants have already developed a certain amount of new root-growth which is easily recognized as being white, against the old root system which would now be mainly brown in colour. Horticulturalists and botanists have already proved that the amount of top growth any plant will develop will depend entirely on the amount of root the plant will, or is allowed to, grow. *Bonzai*, the art of growing forest trees in miniature, proves that anything can be dwarfed by root restriction. Fuchsias are no exception to this rule and a drastic, almost brutal re-potting is necessary to give the new rampant roots room to develop. It is the sight of the newly formed roots which so often frightens the less experienced grower into merely taking a small amount of the old sour compost away for fear of damaging the fragile-looking roots.

What is necessary in the re-potting of fuchsias is that as much of the exhausted compost as possible should be removed from the root ball. In taking such drastic action, it is obvious that many of the current year's roots, which are at a very fragile stage, will come away with the old soil. This is unfortunate but unavoidable and will in no way affect the future of the fuchsia except to delay it for a couple of weeks until the loss is made good. There is enough energy within the storage cells in the stems and hard framework of roots to maintain life within the fuchsia until the new feeding root system is formed.

To remove the old compost it may be necessary to use a pointed stick or a similar tool to dig it away from the established framework of old roots. The fuchsia should then be re-potted into a clean pot of a size definitely no larger than that from which it was originally taken. If when all the old compost is removed only a small number of the old roots remain the plant may well benefit from being put into a smaller pot to aid its quick recovery.

The clean pot should be lightly crocked and the crocks covered with a layer of fresh compost. The plant now held in the position it is to occupy in the pot should have the new compost worked into the remaining root system, and to ensure that no air pockets remain it is a wise policy to bounce the pot gently on the potting bench so that the soil is jarred into any crevices that may exist. The potting complete, the fuchsia should be well watered and put in the warmest, shadiest part of the greenhouse.

No more water should be given to the plants until growth begins again or unless the soil becomes very dry. If a programme of regular overhead spraying is carried out the latter situation should never arise as the spray will supply all the moisture needed by the plant. For the next fortnight, depending on the general conditions, the newly potted fuchsia will appear to stand still. No new growth will appear, although the existing leaves may tend to get slightly larger. While the few leaves on the plant remain turgid, even though there is no visual sign of development, root growth will be taking place within the pot.

As soon as the roots have developed sufficiently, new leaves will begin to appear at the growing points, and the wisdom of the severe action taken at the time of re-potting will soon become evident. The roots now have quite a space to grow and develop, and they will be growing in compost which is fresh, and adequately supplied with nutrients to support healthy top growth. It cannot be over-emphasized that at this stage of the fuchsia's life the correct watering of the plant is of the utmost importance. The ideal state of the compost is to be only slightly moist. If there is a doubt in the grower's mind as to whether the fuchsia needs watering or not, it is much safer to err on the dry side rather than overwater. The little damage that may be caused by the plant becoming too dry can be overcome by giving the plant one good soaking, but the only real remedy for a fuchsia that has become waterlogged is to re-pot into fresh compost once again.

With the warmer weather and the longer hours of daylight that come with May and June, growth will become more rapid and some of the earlier-flowering varieties will begin to show bud even in the unheated greenhouse. From this stage onward the practices are the same as those described for the first-year plants.

We have reached an age when Man is constantly experimenting with ideas to obtain greater benefits from Nature. The fuchsia has not escaped the attentions of the scientists, and already experiments have been carried out with a view to making fuchsias flower continuously throughout the year. Experienced growers will know that even under normal conditions certain fuchsias can bear flowers intermittently throughout the winter months. Papers have been published which prove that by the use of artificial light, flowering can in some varieties be increased a hundred per cent. There appears still to be a great deal of work to be done in this connexion before we can be sure of having fine fuchsias on our tables at Christmas, but it is certainly something that most enthusiasts will look forward to.

It is indeed sad, when so many new and wondrous varieties of fuchsia are being introduced, that so few first-class collections are to be seen. Many of the older collections such as are seen at some of the well-known gardens open to the public continue to show the same varieties for year after year. Such displays are not always a good advertisement for the fuchsia, despite the fact that the plants themselves are often well grown and a credit to their growers, for such collections fail to show the viewing public the versatility and the glorious colour range that there is within the genus.

Everybody has probably a different idea of what shape or form a fuchsia collection should take. Some growers will sing the praises of the 'good old 'uns' while others will discard any variety which is a few years old as old-fashioned. There is no doubt that a collection of variety and major interest should consist of a happy mixture of both old and new varieties. It is a fact that quite a number of the very old varieties that are still available from specialist nurseries have never been surpassed in their class for habit of growth, colour and floriferousness. Such varieties, of which 'Achievement', 'Snowcap' and 'Dollar Princess' are only three, could well form a nucleus around which a comprehensive collection of fuchsias for the greenhouse could be built.

Again, growers must be on their guard against believing that new varieties are automatically improvements on old ones. Since the Second World War the Americans have introduced many new varieties, some of which have proved themselves worthy of inclusion in any collection of fuchsias; but there are unfortunately others which are hardly worth the space they would occupy in the greenhouse. The great range of colour, shape and size of flower given to the fuchsia enthusiast by the hybridizers is there to be exploited to the full.

CHAPTER VI

Training and Growing

There are very few shrubs, if any, as amenable to training as the fuchsia. There are varieties which, by their habit, will force the grower to grow them in one specific way, but again there are many dozens of varieties which can be grown to any number of attractive shapes.

With the exception of hedge fuchsias and the hardy fuchsias in the shrubbery, all should have some form of training. It need be nothing elaborate unless the grower is intending to go in for flower shows. This is particularly true of all pot-grown plants, as, unless some effort is made to train the plants in a greenhouse, they will form a veritable jungle and the fuchsias themselves will become unmanageable. In contrast, if a fuchsia enthusiast wishes to get the maximum benefit from the space available, a little thought to the training aspect of cultivation will allow him to double the capacity of his greenhouse.

Keen gardeners are always short of greenhouse space, and any enlarging of a greenhouse to give plants more room inevitably finishes up with that space being taken up by even more plants, and the position remains 'as you were'. The enthusiast who devotes a greenhouse to the growing of the fuchsia will find that the whole area within that house can be filled. Climbers can cascade their flowers down from the ridge of the greenhouse. Gaps that occur can be taken over by hanging baskets. The staging can be covered by bush-trained plants, with espaliers at the back, and standards and half-standards dotted regularly among them. The front of the staging can have trailing pot-grown plants cascading over the front on to the floor. In fact, there can be a whole world of fuchsia from the floor to the ridge of a greenhouse if only a little attention is given to training.

The fuchsia can be trained principally into the following shapes: bush, table standard, half standard, pyramid, pillar, climber, espalier and hanging basket. There is a variation on the standard which is known as the 'weeping standard' in which a trailing variety is used. Some varieties are better adapted to various methods of training than others, and the Chapters at the end of the book give a list of varieties which are considered to be most suitable for the various shapes described.

BUSH

This is the basic shape of the fuchsia and generally speaking the easiest form of shaping for the majority of species and varieties. Even the training of bush fuchsias can become complicated when exhibiting becomes a serious matter, and the grower goes in for the elaborations required for success. For normal green-house use however the bush is a simple way of training a fuchsia, and should be the first method attempted by the beginner.

The training of a bush plant begins at the rooted cutting stage. The cutting should be allowed to develop four pairs of leaves and at this stage the growing tip should be nipped out between the thumb and forefinger. From this simple operation will grow new shoots from each leaf joint. Generally speaking the four new top branches will be the strongest, and the bottom four will be weaker. This can only be generalized upon as there are some varieties that will send out extremely strong branches from the base, but these varieties are better described under 'Pyramid' training.

Allow the weaker growths to develop unchecked, but the top, stronger branches should be further stopped at three leaves, and these will supply a further four shoots each. For general green-house or outdoor bedding purposes this amount of stopping will suffice, for any more stopping will only delay the flowering of the fuchsia, besides making the plant unwieldy and in need of much training.

Tying for general use will be the looping of the flowering shoots to one central cane. It is better to tie loosely so that the branches arch out and are in the natural fuchsia habit. Too tight tying will only give 'wheatsheaf' effect, and this should be avoided.

Many varieties produce such strong, upright-looking growths that the grower may feel that it is unnecessary to tie them up, but it must be remembered that one branch of a fuchsia often has to support a large number of very heavy flowers at one time, and the first indication one has that the flowers are too heavy for the branches to support is the sight of broken branches littering the staging, or branches so bent that the flowers are rotting on staging shingle.

The only limit to the size of a bush-trained fuchsia is the vigour of the variety being trained and the size of the flowering-size pot, and the latter, for normal greenhouse use and for bedding-out purposes in the first year, rarely exceeds the 48 (5-inch) or 32 (6¼-inch) size.

STANDARDS—TABLE, HALF AND FULL

Once the easy art of growing bush-trained plants has been mastered the grower will undoubtedly turn his attention to the growing of standards. Table standards are primarily grown for those who enjoy taking plants indoors for house decoration. They have a bare stem of not more than 12 inches, and with a neat bushy head look delightful on the hall table or somewhere where they can be well displayed. By this it is not meant that this is the shape to be undertaken by those who grow fuchsias in the house, as the bush is the only practical shape for that purpose; it is more for the grower who in the greenhouse has three or four plants which, by interchanging regularly between house and greenhouse, can maintain a regular display.

The half standard has a straight bare stem of 18 to 30 inches, which is the ideal size for so many amateur greenhouses, the full standard being too lofty in most cases. It is also an ideal subject for summer bedding with bush fuchsias, or any other plants, and gives a little height to what otherwise would be a flat bed. Again when used for this purpose they are not so vulnerable to weather damage as the full standards.

The aristocratic, lofty full standards, which are so admired by the public on the trade stands at Chelsea and other big shows, should have a clear stem of over 30 inches. The only limit to the height over 30 inches is the variety of fuchsia being grown, the skill of the grower, and his ability to house a giant plant. For practical reasons the British Fuchsia Society limit the length of clear stem for exhibition purposes to 42 inches.

The first essential when starting the training of all types of standard fuchsias is a strong, straight, well-rooted cutting. Having obtained this, it should immediately be given cane support. So many growers make the mistake of leaving this support until the stem has reached an approximate height of 6 inches. They are usually deceived by the strength of the cutting and its apparent ability to support itself, but what is not realized is that in any greenhouse, plants are drawn by the light one way or another. This direction-growth may only be slight, but when the stem is eventually tied to something straight a slight bend will be noticed, and this does not smooth out as growth continues but is inclined to accentuate as the stem thickens. So often this slight 'kink' is seen in the stem just above soil level, and this ruins what would otherwise be a perfectly straight stem.

The cutting should be allowed to grow up the cane unchecked. The main essential at this point is that the plant should be kept growing by continually re-potting as soon as the roots reach the outside of the compost. Should the plant become pot-bound, flower buds will certainly appear, and the growth upwards will be checked or even cease altogether. The size pot to aim for is 6¼ inches (32's) for the table standard, 8½ inches (24's) for the half standard, and the full standard will have an even larger pot, or tub, according to its ultimate height and the desired size of the head.

Many growers are confused by the varied instructions given in fuchsia publications concerning the removal of side shoots when the stems of standards are being developed. Some insist that the grower remove all side shoots as they develop, and others say they should remain on the plant, but be kept pinched back, until the ultimate height has been reached. The first school of thought argues that the side shoots take some of the energy from the main stem and therefore slow down its progress to the required height, while the latter school insists that the removal of the side shoots drastically reduces the all-important leaf area, and in consequence upsets the balance of the action of photosynthesis, whereby the plant takes energy from the sun through its leaves, which with the assistance of chlorophyll forms the carbohydrates which are life itself to the plants.

The fact is that there is a purpose in both ways of growing, and whether side shoots are removed or left on the stem depends on the variety being trained. Fuchsias which are capable of sending up stems of great substance, such as 'Snowcap' or 'Melody', will benefit if the side shoots are removed. Nature takes care of the amount of leaf area; the leaves on the main stem, which should remain with the plant until the head of the standard begins to develop, become greatly enlarged by virtue of the fact that there are so few to absorb the moisture brought up by normal root area. If it is the intention of the grower to have a weeping standard or a standard of a lax-growing variety, it is advantageous to allow the side shoots up the stem to develop to two or three pairs of leaves each. This action may well slow down the rate of upward growth of the main stem, but it will help to give it the substance which it would normally be without. The slower upward growth would mean less internodal growth, and the thickening of the stem at each side shoot would be noticeable when the time came to remove those side shoots.

As straightness of stem is one of the hall-marks of good training, ties must be made at 2-inch intervals up the stem throughout the time that it is still soft and still growing up to the required height. Do not tie loosely so as to allow for further development, as too loose ties may cause an undulation in the stem. It would be far better to replace the ties from time to time as the stem thickens. Once the standard is complete, and the main stem has become the thick woody trunk, it would be in order to support the head with loose ties; but never while growth is still soft.

The leading shoot should be stopped when it has made three pairs of leaves above the length of stem which is to remain bare. It is from these six leaf joints that the growth will come to form the head. Training the plant from this stage onwards is similar in method to that of training bush plants. The size of the head will depend on the length of stem on which it is grown. The grower must aim at getting a perfectly balanced head, not so small as to make the plant look all stem, and not top-heavy with an outsize head.

PYRAMID

During the latter part of the nineteenth century, in the age of the large garden, greenhouse and garden staff, the pyramid-shaped fuchsia was considered by many eminent gardeners to be the *pièce de résistance* of plant collections. The huge plants shown by James Lye and others were outstanding for their beauty, and for the amount of work that was put into their creation. It is indeed unfortunate that pyramids are rarely if ever seen these days, for the ordinary enthusiastic amateur has neither the space nor the time to devote to these giants, and the nurseryman has neither the capital nor the labour to spare for them.

The growing of a pyramid should be a challenge to all fuchsia growers, for the only person capable of growing these plants to perfection is one who understands the fuchsia and its cultivation very thoroughly.

Only varieties of strong upright habit should be considered for this type of plant, and any variety which has lax growth, no matter how beautiful the flower, must be rejected. The cutting to start with must be strong, upright, straight and well rooted. The most essential point throughout the whole period of training is to keep the plant growing. Any check, no matter how small, may well wreck the whole sequence and result only in disappointment. This means that the potting on must be done carefully and at the right time. Watering must be sufficient to maintain optimum growth, but never too much. No pest must be allowed to weaken growth, and feeding must be regular once the final potting is done.

Support for the growing plant should consist of one very stout stake in the middle which will be the main support of the whole plant. Round the edge of the pot four long canes should be placed in the compost at equal distances apart and at an angle of 45 degrees. Linking these canes with each other may be lengths of string, raffia or stout thread. This framework is only necessary for the training of the plant, and will be removed before the pyramid is complete.

The cutting should be allowed to grow up to 7–9 inches before the first stopping is made. From this stopping will come at least six side shoots, or if a really suitable variety has been chosen at least eight side shoots. From the two leading shoots the new leader should be chosen. No hard-and-fast rule can be made on this operation, as should a side shoot of much greater substance reveal itself from the next leaf joints down, it would be wiser to rub out the two leading side shoots altogether and take the stronger growth. Whichever growth is selected for the new leader its opposite shoot should be rubbed out. The new leader is now trained up the centre support and the remaining side growths are allowed to develop on to the cane-and-raffia framework around the outside of the pot. When the side shoots have made three pairs of leaves they should be stopped altogether. From this stopping the side shoots should only be allowed to develop two side shoots each, as any more will cause overcrowding and a weakening of growth. The new leader is stopped at four pairs of leaves in order to force further side shoots to develop from these leaf joints. Again a new leader is selected, and the whole process is repeated until the ultimate height the grower has planned for the plant is reached.

The lower side shoots which form the base of the pyramid shape must obviously be the longest of all the side shoots, and it is getting these to the required length which so often baffles the grower. The varieties of fuchsia required to produce the massive proportions of the pyramid shape are negatively geotropic in their habit: that is to say, they have a definite upward growth against the forces of gravity, different from the trailing-type fuchsias which are either diageotropic (horizontal growing) or positively geotropic (obedient to the laws of gravity). For this reason the lower side branches are trained up the temporary framework, for while they are growing upward they will continue to grow strongly, but if allowed to droop or sag they will lose a great deal of energy which will go into those shoots which still point upward. When these shoots have grown long enough the framework can be removed, and the grower will find that the weight of the branches themselves will cause them to fall into position.

The grower who possesses a heated greenhouse has a distinct advantage when growing pyramid-shaped plants, as the easiest method is to grow a summer-taken cutting into a plant to flower in the following spring or summer. This means that slow growth is maintained throughout the winter months. Those without sufficient heat must start their training as early in spring as possible so that at least the shaping can be completed by the time growth stops. During the winter the dormant plant should be rested on its side and turned at very regular intervals. This action will give the lower buds a distinct advantage, as the natural inclination of the sap to go directly upwards is arrested by the fact that the main stem is horizontal instead of vertical. When the growing season begins once more, the pyramid will naturally be stood upright and the sap will quickly transfer its main flow to the apical buds; but by this time the lower side growths should already be breaking into growth. A pyramid can be maintained in this shape for three or four years, after which it becomes difficult because the fuchsia will make the bulk of its new growth at the top of the plant and ruin the gentle taper from the top to the bottom. Winter care of established plants will entail the laying of the fuchsias on their side for reasons already given, and pruning each spring will consist of gentle pruning of the bottom growths with more drastic action at the top.

During their heyday pyramids were extensively used by head gardeners to line the sides of paths leading to the master's house, where their stature and grandeur did much to impress visiting dignitaries.

PILLAR

An alternative shape to the pyramid in that golden age of fuchsia growing was the 'pillar'-, sometimes described as the 'cordon'-, trained fuchsia. Whereas the pyramid gave a 'Christmas tree' impression, the pillar is completely devoid of any taper; the object is to maintain a straight column of flowers about 18 inches wide from the pot to the top of the plant. The varieties to grow into this form should be identical in vigour to those used for the pyramids.

There are two methods of producing this shape of fuchsia,

both involving exactly the same principles, but in one case the result is achieved by putting two plants together in one pot; and the other method uses one cutting only. The two-plant method is easier and quicker in producing results, and in this age of speed would perhaps be the more popular one.

Two vigorous cuttings are placed 1 inch apart in the same pot and to simplify the explanation of their training they will be called cuttings *a* and *b*. Assuming that the 'pillar' is to have a finished height of 6 feet then *a* will supply the flower to the lower 3 feet and *b* the flower from 3 feet to the top. Both *a* and *b* will be potted on together and can have the same supporting stake if it is strong enough. In all aspects of their culture they should be treated as one plant and only separated when their points of training are referred to. Cutting *a* is allowed to grow upright for a distance of 3 feet, still assuming the finished pillar to be 6 feet, and is then stopped. All resulting side shoots are pinched so that they are even in length. In the meantime, cutting *b* should be growing in standard fashion with its side shoots being removed to just beyond the 3-foot height of cutting *a*. From that height the side shoots are left on, and when *b* is stopped at 6 feet these will develop to form the top half of the plant.

To those who have never witnessed the growing of plants trained in this fashion it may sound as though the bottom half would flower well before the top half. In practice the check that *a* gets when it is stopped is enough to hold it until *b* reaches its maximum height. Should one cutting prove to be slightly less vigorous than the other it should automatically be selected as *a*.

The pillar made from one plant is similar in training except that the single cutting is stopped at three pairs of leaves and the two strongest side shoots are selected to represent *a* and *b*.

Growers new to fuchsia growing may wonder what advantages the two-plant method has over the single-plant method and vice versa. It should be pointed out that the two-plant 'pillar' will never be accepted for exhibition at shows under the present rules governing the showing of fuchsias. On the other hand, for general garden use the two-plant pillar produces much quicker results, having avoided the initial check at the three-pair-of-leaves stage. Furthermore, after a period of winter

Training and Growing

dormancy the resumption of even growth is much more reliable in the two-plant than in the single-plant pillar.

GREENHOUSE CLIMBER

Many people, and keen gardeners among them, can only imagine the fuchsia as a shrubby plant. Among the species, however, are many different kinds which in their natural habitat scramble over rocks and through other shrubs and trees, sometimes climbing to quite a height in their search for sunlight. While this characteristic habit of growth is not always evident in the hybrids, it can often be brought to the fore by careful training. Before the rebuilding of the No. 4 Temperate House at Kew Gardens thousands of people admired climbing plants of varieties 'Royal Purple' and 'Rose of Castile Improved'. Many may remember the magnificent climbers shown by L. R. Russell Limited on their 'Gold Medal' stands at the Chelsea Flower Show when the variety 'Muriel' was shown to perfection trained in this way.

Many people feel that the true beauty of the fuchsia is not evident unless the plant is viewed from beneath. The exquisite beauty of colour and form of the corolla can be much better appreciated in this way. The varieties to be chosen for the purpose of training a climber must be vigorous in their habit of growth although they need not be of upright-growing type. The variety 'Muriel' which has proved so successful at the Chelsea Flower Show is a vigorous grower although its habit of growth is almost cascade. 'Royal Purple', grown so well at Kew Gardens, is however a strong upright grower. For the average amateur greenhouse owner it would perhaps be better for the vigorous cascade type to be grown rather than the upright type, as it would be easier to confine its growth to the limits of the small greenhouse. A strong cutting should be selected and grown on as though the plant is to be a very small standard. All side shoots should be removed until the fuchsia has reached the point between the eaves and the ridge; and from this point they should be left on the plant. Any loss of vigour in the leading shoot must result in the leader being pinched out and the strongest side shoot taking its place as leader. Nothing must be allowed to stop the plant reaching the required height quickly.

There are few conventional pots of the size to maintain a climber for any length of time, so it would be better if the plant were planted directly into the greenhouse border; or, if this is not possible, into a specially built-up container which would give the plant plenty of root room and allow the grower to re-move the depth of old soil every spring and replace it with a top dressing of fresh compost. When the leader has reached the required height it should be pinched out and the main shoots should be tied to wires arranged between the eaves and the ridge of the greenhouse. The wire can be that sold in horticultural shops as 'garden wire'. Only the main shoots should be tied up and the secondary branches should be allowed to cascade down to give the best effect.

Maintenance in subsequent years will consist, besides the top dressing already mentioned, of a pruning of the secondary growth to get a reasonable covering of the roof. Those growers who use permanent shading such as 'Carsolumbra' or 'Summer Cloud' must make sure that it is applied to the outside glass reasonably early in the year to prevent the young shoots scorching in the spring sun.

Espaliers

The espalier-trained fuchsia has been neglected for far too long. It has a very definite purpose and has the virtue of showing off many varieties of lovely fuchsias to their best advantage. Practically any variety of fuchsia can be trained in this way, but it is particularly useful in displaying some of the trailing varieties whose beauty of flower can only be appreciated at eye level. In a greenhouse of fuchsias their place would be at the back of the staging against the glass, hiding the bare stems of any climbers the grower might have, and taking up space that might otherwise be wasted. Unless this shape of fuchsia is required for exhibition purposes it is better if the plant is grown and trained *in situ*. Grown individually so that they can be moved around they are much inclined to be top-heavy, besides requiring some elaborate form of framework. On the other hand if a number are grown in permanent positions at the rear of the staging, wires fixed to go the length of the greenhouse will be the only equipment needed.

As with all trained plants the grower should ensure that the cutting selected has been well rooted and is of good stock. This should be grown on in a large 60 pot with just one stake supporting the central leading shoot. As soon as the roots are moving strongly around the outside of the compost it should be planted directly into a 32 pot (6¼ inches). By normal practice this would be over-potting, but if the grower plants the soil ball deeply in the new pot and does not overwater there will be no trouble at all. The whole object of this move is to get the plant into its flowering-size pot as early as possible so that the plant can be put into the place it is to occupy for the whole of its flowering period. Training of the plant to espalier can now begin in earnest.

The horizontal wires along which the lateral shoots should be trained should be 6 inches apart. A cane should be tied to the wire in each case where the leader is to go, so that it can have support for all its length. The side shoots or laterals at the leaf joints which correspond with the wire must be allowed to develop. The laterals that start to develop at places between the wire supports should be rubbed out as soon as they appear so that the whole energy of the plant goes into the wanted growths. The grower must watch for any exception to this rule and give a little thought before actually pinching out the apparently unwanted shoots. For example, the laterals at the required heights may not be developing strongly enough, in which case the leader should be removed just above a lateral which would normally be rubbed out, and the lateral trained to be the new leader. This check to the top growth will result in renewed vigour being passed to the required laterals. Again, if the leading shoot shows signs of diminishing vigour, it should be removed and a lateral again taken as a new leader. Sometimes if the side shoots show signs of slowing up their rate of growth before the required length is reached new life can be given to them by training the laterals upwards, and when they are long enough tying them down to the correct shape.

The laterals and the leader should be stopped when they are as long as the grower requires them, or at a time at least seven weeks before they are required to flower. Once the laterals are stopped, shorter secondary shoots will appear complete with flower buds, and when the grower has, with experiment, learned the characteristics of the varieties being grown, and his timing is perfect, he will find that he can produce a veritable wall of colour.

During the winter the plants should be rested, still supported by the wires along which the laterals are trained. This is necessary to prevent damage being inflicted on the plants by over-handling. When growth begins once more in the spring they should be taken down for re-potting and pruning and then immediately returned to their place. Pruning will consist of cutting the laterals back to two joints from the main stem and selecting one shoot from each for training in the same way as the old lateral the year previously.

HANGING BASKET

This is thought by many people to be the most effective way of displaying the fuchsia. Truly the wonderful hanging baskets to be seen at Kew Gardens and at the annual shows of the British Fuchsia Society are good arguments in favour. The Americans love this type of fuchsia, and besides growing it in the conventional wire basket they enjoy having novel and decorative containers in which to hang the plants. Furthermore it is from the United States of America, particularly the state of California, that we have the vast majority of the trailing type of fuchsia so necessary to basket culture. Tradition, in Great Britain, dies hard and there are few growers who would dare to grow their trailing fuchsia in anything but the traditional wire basket. This is perhaps a pity, because once a lead was given by somebody whose name was a household word to the gardening public, there are many who would be overjoyed to follow suit, and a new pleasure, the pleasure of being able to choose from an unlimited range of containers, would be added to an already happy pastime.

The growing of basket plants starts at the cutting stage. When taking the cuttings the grower must ensure that he will have enough for the number of baskets he wishes to fill. The 15-inch wire basket will take four plants of varieties such as 'Cascade', 'White Spider' and 'Niobe', but may need five plants of such weaker-growing types as 'Enchanted' and 'Pastel'. Again, if growing for exhibition care must be taken as to the number of

plants allowed in a single basket. The cuttings should be grown
on as ordinary pot plants until they have a thoroughly strong
and active root run. If the varieties being grown are of the full
cascade type they may well benefit from a little upright support
at this stage; but those which are semi-cascade, such as
'Marinka', can be left to grow as they will.

While the rooted plants are growing thus attention must be
paid to the preparation of the basket itself so that before they
become root-bound the plants can be planted directly into their
flowering position. Basket preparation is an art in itself and
must be undertaken with great care and forethought. It requires no skill to grow the weak efforts so often seen growing in
the porches of suburban houses and elsewhere, and if there was
a Society for the Prevention of Cruelty to Plants most of their
owners would face prosecution. The skill in growing those
magnificent baskets seen at Kew and in some public gardens
does lie in the preparation of the basket, so it is as well that it
should be discussed in some detail.

In the past the most difficult part of basket culture has been
the watering. As the baskets are suspended in space completely
surrounded by air it is obvious that during a hot dry period,
when the rate of respiration of the plants themselves is very
high, a very great deal of moisture is removed direct from the
compost by evaporation. This has often meant to the grower,
who has to earn his living away from his plants, that someone
else has had to be approached to watch the baskets during hot,
dry weather. This difficulty can now largely be overcome by
lining the inside of the basket with polythene and leaving a small
hole at the bottom for drainage purposes. The only objection
raised against the use of polythene is that it is unsightly compared with the old rustic appearance given by packed moss;
but there again, if the plants are grown properly not only will
the polythene become hidden but the whole structure of the
basket will disappear beneath a mass of fuchsias. The usual
moss lining should be placed over the polythene as in normal
basket culture, as this will prevent the sharp bits that may be
present in the compost from penetrating the polythene and
making it too porous. Furthermore the moss should be packed
to stand above the edge of the basket so that the water does not
just flow off the surface when the basket is being watered.

The basket is now ready to receive the plants. A layer of J.I. Compost No. 2 should be placed over the bottom of the basket and the four plants taken from their pots and laid at an angle of approximately 45 degrees on top of it. If the plants have been trained upright while in their pots, this angle will now ensure that the growth will begin to suspend some little distance from the edge of the basket, and thereby give the impression that the basket has a greater circumference than it actually has. The transference of the plants from pots to basket should be done with as little root disturbance as possible, as any check in growth at this stage will have some effect on the ultimate result.

J.I. Compost No. 2 should now be worked between the four root-held balls of compost. It is essential for strong even growth that no air pockets are left within the baskets. This can be prevented by working the compost a little at a time among the pot-shaped balls and making sure that it is firm before any more is added. The roots being finally covered, a saucer-shaped hollow should be left in the centre of the top of the basket, so that it can be watered with ease. Some growers have a little difficulty in maintaining this depression, as after one or two waterings the soil becomes flat again. To overcome this and to ensure that the maximum water reaches the soil within the basket it is often a wise move to sink in the centre of the basket a 3-inch pot and round it, some 3 or 4 inches away, four 'thumb' pots. Together with the rim of moss around the edge of the basket this will ensure that all parts of the compost receive their full quota of water when the operation of watering is carried out.

For many years those growing varieties for baskets such as 'Marinka' and 'Mrs. Marshall' had no difficulty in growing their baskets in a neat open order, for such varieties, although they can be made to cascade, have a certain stiffness of stem which does prevent all the blooms touching one another as they hang down. The modern American basket varieties such as 'Cascade', etc., so desirable because of their vigour, the size and colour of flowers and their form, are so lax in their habit of growth that if allowed just to hang they will do just that; and because of this all the branches will hang down together and the flowers will touch. This will cause the flowers to bruise each other as the baskets swing in the breeze or wind. It is obvious therefore that such varieties should have some framework on

94

the basket itself to open out the growths. A number of canes pushed into the basket so that they protrude some inches over the edge, joined together by a network of string or raffia, will give the basket a much lighter and neater effect if the branches are trained away from the basket at different distances along the proposed framework.

So far what has been written applies only to the hybrids, as the training of species in any way but their natural form does far more harm than good. They do, however, vary considerably in themselves in their habits of growth, and practically all the shapes mentioned, especially bush, basket and climber, are normal types of growth within the species. Chapter XI gives details of the commoner species which benefit from the exploiting of their natural habits.

CHAPTER VII

Pests and Diseases

The fuchsia is, by modern standards, a shrub which is comparatively free of any dread disease, and has only a few pests to contend with. Other plants and shrubs which have reached the same stage of high development as the fuchsia have developed also a constitution which is susceptible to numerous viruses and functional disorders. Hybridizers of fuchsias of the future must guard against any overdevelopment which might cause any breakdown in the constitution of the shrub. Besides its natural beauty, the ease of growth and the freedom from disease are the two greatest assets the fuchsia has to offer the grower, and it must be seen to that it keeps these virtues for all time.

Before dealing in detail with the specific pests and diseases that attack the fuchsia, it must be said that whatever cures are achieved with the pesticides and fungicides available today, the best plants will always be those which have never been weakened by such attacks. A plant stunted and disfigured by a severe attack of aphis will never attain the size and floriferousness expected of it, even when all the green fly are dead. There is no substitute for prevention, and only a regular routine of spraying outdoor-grown fuchsias, and the regular greenhouse treatment, will ensure that plants are free from pests at all times.

PESTS

Aphids, Green Fly and Black Fly

The green fly is perhaps the commonest pest attacking the fuchsia both in the open and in the greenhouse. Black-fly

96

attacks are not so common, but as the results of the attacks are identical they are treated here as one pest. They damage and weaken the plants by sucking out the sap needed for growth. Very heavy infestation will cause stunted growth and even slight infestation will cause 'leaf curl' and 'leaf drop', sometimes to a considerable extent. Ants encourage the formation of aphis colonies as they 'milk' the aphides for the honeydew that they give off. It is therefore essential that any ant nests in the vicinity of the fuchsias be wiped out by a dusting with B.H.C. dust; it is useless to kill off the aphis if, within a few days, the ants reintroduce them to the plants.

The normal cultivation for fuchsias, which should include a daily overhead spray with clear water, will definitely reduce the attacks of green fly; but instead of green and black fly occupying their normal attacking positions at the growing tip they will take up defensive positions underneath the lower leaves. Here they are free from attack unless the spray lance is used to spray from the bottom up into the plant. When the lower leaves of the fuchsia turn yellow and begin to drop off they should be inspected for green fly, as quite a mild attack may have this effect on some varieties.

The strength of a green-fly attack is the ability of the pest to multiply rapidly. Two or three aphides on a plant one day may well mean many thousands within a few days. Because of this it is essential that the spraying programme be one of prevention rather than one of cure. The insecticides should be sprayed at regular 10-day intervals rather than when green fly are seen on the plants, as only by this means can it be assured that no damage is sustained.

Systemic insecticides work thus: the fuchsias are watered with the insecticide and the roots take up the solution, thereby killing the pests as they feed; these are particularly effective on fuchsias. For fuchsias grown in the open this method of prevention is far too expensive and extremely wasteful, but for pot plants it is ideal. For those who prefer the normal spraying insecticides there are numerous different brand names on the market. Most of the modern sprays contain such powerful agents as lindane, benzene hexachloride (B.H.C.), and malathion. The older types of agent such as nicotine, derris and pyrethrum are produced for more specific purposes these days rather than as general insecticides.

GREEN CAPSID BUG (*Lygus pubulinus* L.)

This is a bright green insect not unlike an overblown aphis which is extremely active and travels rapidly round a large number of plants. The damage it causes can be readily seen on fuchsias, usually in spring or early summer. It causes this damage by sucking the sap from the growing tips, but before doing so it injects into the plant its own form of saliva in much the same manner as the mosquito acts on the human being, and it is this which causes the growing tips to distort and blister and very often go 'blind'.

The green capsid bug seems to confine its activities to the outdoor-grown plants and those grown in the cool greenhouse. While it is very unlikely that the damage done by capsid bugs will cripple the plants it is certainly unsightly and very often ruins all efforts to obtain a shapely plant.

This pest can be quickly eliminated if the fuchsias are sprayed with B.H.C., a D.D.T. emulsion, or lindane. If the normal spray routine is carried out or if systemic insecticides are used at regular intervals it is probable that the grower will never know the presence of the pest.

The green capsid bug has a very near relation in the tarnished capsid bug which is slightly larger and of a reddish-brown colour. Attacks by this bug on fuchsias are not so common and their control is the same as for the green capsid bug.

WHITE FLY (*Trialeurodes vaporariorum*)

This is the pest which really threatened the continuance of the fuchsia as a garden plant. There is no doubt that the tomato has had the greatest impact on the vegetable market since the potato was introduced into this country, and because of the many thousands of pounds spent on its development and cultivation, anything which in any way threatened its well-being was to be quickly eliminated. The white fly attacked an unlimited number of greenhouse plants but it certainly did seem at one time that it had chosen as its host plants the tomato and

Alice Hoffman

Billy Green

Blue Waves

Bon Accorde

Bridesmaid

Cardinal Farges

Cascade

Checkerboard

Constance

Dorothea Flower

Dr Topinard

Emile de Wildeman (Fascination)

Flirtation Waltz

Liebreitz

Marin Glow

Miss California

the fuchsia. At the time when this was most evident, at the beginning of the First World War, there was no real means of defence against the pest and many believed that it was in fact the fuchsia which was encouraging the white fly to attack the tomato. Experiments have since proved that a neglected tomato will still suffer white-fly attack even when there are no fuchsias to act as host.

It was the introduction to commercial growing of the chalcid wasp (*Encarsia formosa*) which at last gave some measure of control. The scale stage of the white fly is parasitized by the wasp and the scales attacked turn black and die. Provided the attack has not got out of hand this method of control is still highly effective and the parasites are still obtainable from the Greenhouse Research Station, Littlehampton, Sussex.

While the greenhouse grower who methodically sprays and fumigates his greenhouse may go for years without seeing a white fly, any slight lapse in the programme will quickly show the susceptibility of the fuchsia to this pest. The first sign of the presence will probably be the flight of the adults when a plant is disturbed, and it is essential to the well-being of the whole greenhouse that measures be taken immediately to eliminate the pest. Severe damage is done to the plants by the nymphs, the young flies sucking the juices from the leaves, and the honeydew they secrete blocking up the breathing pores of the plant. The pests multiply at a great rate, and large numbers of plants can be ruined in a very short space of time.

It is fortunate that a large number of the modern insecticides are capable of destroying this pest, including D.D.T., malathion, and B.H.C., and these can be either used in spray form or as smokes (fumigants).

CYCLAMEN MITE

It is only in recent years that this pest has been found to attack fuchsias to any extent. Its attacks would appear to be becoming more and more frequent, mainly because most growers are unfamiliar with the damage sustained by the attacked fuchsia, and again not enough publicity is given to the pest. Normally this is a pest which enjoys the dry conditions which are the opposite to those enjoyed by well-grown fuchsias, but the pest has found an entrance via the back door. The attacks on the plants appear to start when the plants are being dried off for winter storage and the effects are therefore not noticed until the fuchsias are started into growth in the spring. The growths of the attacked plants are thin and stunted and the leaves are small and have rust-coloured patches on them.

Should dry conditions occur during the growing season it is quite possible that the pest will attack the plants in full growth, when the first signs will be the malformation of the newer growths and the failure of the flowers to open. The flower buds burst in some cases but the sepals fail to open; and in other cases the buds reach maturity and then turn brown without opening at all.

If a fuchsia is seriously attacked by this pest it is cheaper and more convenient to burn the plant. However, if it is a plant that for some reason or another the grower must keep, or again if the attack has not reached serious proportions, the following procedure should be adopted. Remove the plant from its pot and scrape and scratch from the 'root ball' as much of the old soil as possible. Soak the denuded roots in a solution of B.H.C. or malathion, making sure that the solution reaches all parts of the 'soil-ball'. Re-pot the fuchsia in fresh sterilized compost but if possible in a size smaller pot than that from which it was taken originally. This will enable the plant to make a quicker 'get-away' and therefore minimize the damage already sustained. Fumigate the greenhouse in case the pest is still present.

Red Spider

This mite will attack and devastate fuchsias in the greenhouse or frame but rarely will it attack those plants grown in the open. Officially this should never be listed under pests that attack fuchsias, for it only attacks the plants in conditions under which no fuchsia should grow. The red spider mite will only flourish under dry conditions, and the grower must look to the conditions under which he is growing the plants rather than rely on modern chemicals to provide the cure. Unfortunately there are many growers of fuchsias who get these attacks from the red spider, which only proves that the plants are not being allowed to grow as well as they should. Furthermore, great damage is done to fuchsias by this pest, for once they are established in a greenhouse they are extremely difficult to eliminate.

The microscopic mites breed with great rapidity, quickly covering the underside of the leaf. The attacked leaves turn brown in colour and drop from the plant. Once an attack is detected it will not be long before a whole plant is defoliated. A quick counter-attack by the grower may well save the plants, but according to the severity of the attack, so will the fuchsias be weakened. It is essential that once red spider has been detected on one plant it must be assumed that all neighbouring plants are affected, and must be treated immediately. In fact the whole greenhouse must receive attention, so that the pest is quickly eradicated.

Modern science has provided many cures for red spider, but the most effective are the azobenzene smokes or the greenhouse aerosol. The manufacturers of the insecticides give detailed instructions as to their use, and these should be followed closely in order to get full benefit from them.

The grower, having rid the greenhouse of the pest, must ensure that conditions are such as to resist a further attack. The greenhouse must be shaded from hot sunshine and the plants must be grown in that cool damp atmosphere they like so much. Up to the time the flower buds appear the plants can be vigorously sprayed every day either with the spray or with the hose. With the arrival on the plant of the flowers the spray should be directed at the pots themselves, the staging and the paths of the greenhouse. At no time should the fuchsias themselves want for water.

Leaf Hopper

This is not a serious pest of the fuchsia, but since its attack can make many a plant unsightly it is necessary to prevent this insect coming in contact with the fuchsias. Generally this pest attacks plants in the open, but it is often seen on fuchsias in cool or cold greenhouses.

This is a very active, pale green, jumping and flying insect, and the adult hoppers take small flying leaps into the air when the plants are disturbed. The most serious attacks on fuchsias occur in the spring, when the wingless, colourless nymphs suck the sap from the leaves, which in turn gives the leaves a mottled or speckled appearance. A glance under the attacked leaves will often reveal the presence of their cast or moult skins, and these are sometimes taken for the nymphs themselves.

All the modern contact sprays are effective against this pest, and if the recommended spraying programme is carried out this pest will present absolutely no problem to the grower.

THRIPS

This pest can prove very serious both under glass and in the open garden. Like the red spider mite it flourishes in dry arid conditions, and although it would be unusual for a fuchsia to die as a result of the attack of thrips, the attack would quickly ruin the plant's decorative value. With the control the grower should have in the greenhouse this pest should never be allowed to disfigure plants housed therein. However, the open garden does present a bigger problem as conditions are not always under the control of the grower.

The adult insects are slender, with a narrow body seldom longer than one-tenth of an inch, and are dark brown in colour. They feed by breaking up the tissues and sucking the mashed foods through their mouths. Both flowers and leaves are attacked by this pest, and as already stated the appearance of the plants after the attack is far from decorative.

Timely spraying of D.D.T. or pyrethrum spraying extract will prevent attacks on fuchsias by this pest in the open; in the greenhouse D.D.T. fumigants, nicotine shreds, or the use of the greenhouse aerosol will quickly destroy the pests.

CATERPILLARS

Fuchsias grown in the open garden are often attacked by the tiny caterpillars of one of the *Tortrix* moths. They feed under the leaves, eating away the succulent part of the leaf leaving just the upper epidermis, and a quick glance gives the impression of a scorched leaf. They do not endanger the life of the attacked fuchsia, but as they drop from leaf to leaf on their silken threads they can spoil quickly the appearance of the plant.

All the general contact insecticides will kill these caterpillars, which really means that there is no need to take any special action if a regular spraying programme is carried out.

VINE WEEVIL

It is not known that the adult weevil does any damage to the fuchsia, but the creamy-white, brown-headed, legless larvae feed below ground and have been known to do considerable damage to the root systems of the plants, sometimes causing the complete collapse of the fuchsia. As always, prevention is better than cure, and it is therefore essential that the greenhouse be cleared of the adult weevil that is responsible for the larvae being there. These black beetle-like insects hide among the debris and litter in the greenhouse, so the first duty of the grower if he wishes to prevent an attack by the pest is to clear all rubbish away and wash the inside of the house, prior to the start of the growing year, with a solution of Jeyes Fluid.

The re-potting of the plants will reveal the pests in the soil, although if an attack is suspected and no sign of the pest is found during this operation the roots should be soaked for a few minutes in a solution of liquid malathion.

These then are the pests which regularly attack the fuchsia. From time to time other insects will, in passing, 'sharpen their teeth' on some part of the plants. The leaf-cutting bee and the caterpillar of the elephant hawk moth may well take lumps out of a leaf here and there, but it is only for the want of something better, and they are soon on their way. Wasps may suddenly launch an attack and remove the style and anther from every flower, but if the nest is not on your ground there is little you can do to stop their ravages except to kill wasps that happen to be handy when you are carrying out your routine spray or fumigation. Such pests, however, do not single out the fuchsia as their main object of attack and must be treated as general rather than specific pests.

DISEASES

Despite the inbreeding and crossbreeding that have gone into the hybrid fuchsia over the past two centuries, there are no diseases or functional disorders which seriously affect the growing of the shrub. In point of fact there is no known disorder brought about by the actual act of hybridizing, and it can be said that in many cases the constitution of the fuchsia has improved, for the majority of hybrids are much more tolerant of mishandling than many of the species.

BOTRYTIS CINEREA

This fungus has been known to attack growing fuchsias, when usually a new shoot will turn black about half-way up the branch and all the growth beyond that point will collapse. This is only known to attack unripened wood, and a close inspection of the blackened portion will reveal the fine hairs which are the fungus structure. However, the most devastating attack is when the fuchsia is in its dormant condition during the winter months, when the fungi will kill all the growth buds on the ripened wood. While the plants look healthy, they will produce no shoots in the spring owing to the attack of this fungus and many growers of fuchsias have been puzzled as to why an apparently healthy plant should fail to grow when the time and conditions are conducive to growth. Sometimes the attacked plants will produce growth from below soil level, but even in these cases the shape and vigour of the plant suffer so that a whole season is needed to replace what has been lost.

The fungi flourish and spread when conditions in the greenhouse are damp, cold and stagnant. This means that although a grower cannot always supply the necessary heat he must always supply a free circulation of fresh air. Watering during autumn and winter should be given special attention. It is always better, although not always possible, when giving fuchsias the few waterings they require in winter to take the plants outside the greenhouse and soak them and allow them to drain while still outside, so that the excess water is not allowed to lie in the greenhouse and so encourage the fungus.

The remedy is not difficult with all the modern aids available. First action must be to remove all parts of a fuchsia affected by the fungus. Spray the whole greenhouse and everything it contains with a solution of Captan Fungicide. It is also a point to remember that the tools used to cut away any affected part should be washed in that same solution.

Botrytis cinerea is not a difficult fungus to rid oneself of, but if it is allowed to go unchecked it can inflict considerable damage on a collection of fuchsias.

LEAF DROP

This is neither a disease nor a functional disorder, but Nature's way of telling the grower that he or she is doing something that the fuchsias object to. The leaves turn yellow and drop off the plant. This condition can be brought about by aphis attack, but these pests can be seen on the fallen leaf, so it is for those leaves that yellow and drop for no apparent reason that an explanation must be given.

There are three reasons for this malady and they are

(*a*) excessive watering,
(*b*) lack of water,
(*c*) excessive use of fertilizers.

It is the first two of these cultural faults that are the most common, as the fuchsia is a gross feeder and it is difficult to give too much feed. At no time during its growing period should the fuchsia be allowed to dry out, for when water is given to a dry plant the moisture is immediately taken up by the roots and the pressure on the cells of the limp leaves is such as to burst them, causing the death of the leaf.

Excessive watering usually takes place before the fuchsia reaches its pot-bound state and is an extremely dangerous condition, not only for the appearance of the plant but for its very existence. The lack of oxygen caused by the sodden state of the compost kills the fine root hairs, which die back and cause in time the death of the whole root. The remedy is to remove the fuchsia from the pot and re-pot the plant in fresh compost in a size smaller pot. By doing this the roots quickly reach the outside of the smaller pot and their growth aids the recovery of the plant.

CHAPTER VIII

Propagation

Many plants maintain a certain 'snob' value because they are difficult to propagate. Orchids up to quite recently were only grown by the relatively moneyed people because in the first place they were expensive to produce. Camellias are grown by far too few people solely because being shy to make root they become expensive, and therefore exclusive. Fuchsias on the other hand are a flower for the masses. The ease with which they can be propagated means that quite large collections can be acquired for quite a modest sum of money. The fact that most up-to-date and outstanding varieties can be purchased for such a meagre amount may tend to make some of the gardening fraternity look down their noses a little, but to the majority who grow only those plants that give the greatest return for the money they are prepared to give, the fuchsia is rated very highly indeed.

There are only two recognized ways to propagate the fuchsia; by cuttings and by seeds. It can truthfully be said that the greatest percentage of gardeners, both professional and amateur, have rooted a fuchsia cutting at some time during their gardening life, but there are few who have raised them from seed. Fuchsias have been successfully grafted in the past but nothing has been achieved by this method of propagation, except that a little time has been wasted by someone who has time to waste. Anything that is achieved by grafting can be achieved more quickly and easily by cuttings, so there seems little point in continuing with this method.

Propagation by seed is Nature's main way of perpetuating the species. Nature has applied to this operation all the wonders that are life itself. Hybridizers must of course use this way when raising new varieties, and seeds of the species are often sown by enthusiasts for the purpose of reproduction. For those growers who have no desire to gather their own seed, there are several nurserymen who offer seeds of the species and mixed hybrids, but it must be said that there is much doubt and a lot of argument over how long the fuchsia seed remains in tip-top condition. The majority of successful hybridizers emphasize that fuchsia seed should be sown as soon as possible after it is gathered, and the results that they have achieved give great force to this argument. This does not mean that a packet of seed purchased from a first-class nurseryman will not produce some results, but little if anything of note has ever come from such effort.

Cuttings are the method adopted by growers to increase or renew their stocks of fuchsias. The fuchsia roots so easily in so many ways, and in so many rooting media, that all growers develop their own particular way of performing the operation; much to the dismay of the beginner, who wonders exactly who is right. However, there are rights and wrongs in rooting fuchsia cuttings and these will be dealt with later in this chapter.

SEED PROPAGATION

The gathering of the fuchsia seed is detailed in full in the chapter on 'Hybridization'. Sowing, as already mentioned, should take place as soon as possible after the seed has ripened, but in the event of this being impossible they should be placed in an airtight container of some kind and stored in a cool place until the grower is ready to proceed.

The time of year most suited to seed sowing will depend on the amount of heat the grower can muster at any time. The ideal temperatures for rapid and healthy germination are a 65°F. maximum and a 55°F. minimum. Some people will argue that a variation of 10° is too much for anything as delicate as a fuchsia seed, but this is not so. Only the grower who has his greenhouses or frames under constant surveillance can hope to maintain an even temperature, and there are few such people about. Variations in temperature may be even more than those stated without serious damage to the resultant seedlings, but the more even the temperature then the more even will be the germination.

The importance of the operation of seed sowing cannot be overstressed, especially in the case of the hybridizer. A whole season's work can be ruined by a little unnecessary carelessness, impatience or lack of attention to detail. If the seed should get away to a poor start, the poor root system and growth will immediately give the hybridizer the idea that the plant is of inferior breeding, and only fit for the bonfire, whereas that same seed given the ideal conditions may prove to be first class and of robust constitution.

Initially, the most important thing is cleanliness. A germinating seed is at this period most vulnerable to any attack that may be made upon it, whether it be fungus or insect attack. Nature, decreeing that only the fittest survive, provides but slight natural protection to the young of anything, whether it be beast or vegetable, and in the case of seedlings the grower must provide the necessary protection.

The receptacles in which the seed can be sown can either be earthenware seed pans, alpine pots, ordinary 3-inch clay pots, or the conventional wooden seed boxes. In the case of the earthenware or clay receptacles, they should be scrubbed prior to being used, and rinsed in a weak solution of Jeyes Fluid. Wooden seed boxes should be treated in the same way but should also be given a coat of preservative such as Cuprinol Green, or Solignum VDK, some few days before they are required for use.

Many growers have their own pet media for the sowing of seed, as they have for many other aspects of growing, but the one medium which is easy to acquire and which has proved so generally successful is John Innes Seed Compost. Purchased from a reliable source, or made up by the grower, using the correct ingredients, it supplies the dormant seeds with everything needed to start, and maintain, a healthy life.

The formula for J.I. Seed Compost, which is the result of research work at the John Innes Horticultural Institution, is as follows:

Loam (sterilized)	2 parts (by loose bulk)
Peat (high grade horticultural)	1 part (by loose bulk)
Gritty sand (clean)	1 part (by loose bulk)

To each bushel of this mixture is added:

1½ oz. superphosphate of lime
¾ oz. ground limestone or chalk.

The whole of the ingredients are passed through a ⅜-inch mesh sieve.

Probably the most important part of this formula is that the loam used is sterilized. Having taken such pains to ensure that the sowing receptacle is clean, it would be useless using a medium that contained an unlimited number of harmful bacteria, soil pests and weed seeds. Furthermore there is nothing more exasperating than to have a vast number of differently shaped cotyledons appearing from the soil and not knowing which are those of the fuchsias.

Unless the grower is to sow a great number of seeds there is no doubt that it is more economical and practical to purchase the compost from a respected horticultural sundriesman. Such a person realizes that the success of his business can only be measured in the number of repeat orders that are received by him, and for this reason will only give of his best.

Before placing the compost in the sowing receptacle, the question of drainage must be understood and its principles applied. At no time during the life of a fuchsia plant is drainage more important than at the seed-sowing stage. The finest material for use as drainage is still what gardeners call 'crocks'. This is merely broken pieces of flowerpot or household china further broken to a convenient size for the job in hand. It is known that some growers use peat or broken turf as drainage material, and to stop the loss of compost through the drainage holes at the bottom of the receptacle, but this is not recommended in the case of fuchsia growing, and especially when the life of some precious seed is at stake. The amount of material to use for drainage purposes is always a problem to the beginner, who is usually worried in case he is doing anything wrong. Whatever the shape or size of the vessel into which the fuchsia seed is to be sown, it should contain one-third drainage material covered by two-thirds seed compost. This may seem to many to be a little on the generous side as regards drainage, but anything less will often result, at the first good watering, in the finer soils sifting through insufficient 'crocks' and interfering with the draining of excess water when the subsequent waterings are given.

The condition of the compost at sowing time is also of great importance. It should be neither too wet nor too dry, and the recognized test, which is an ideal guide to the grower, is to take one handful of compost and compress it with that same hand, open the hand, and the compost should remain in a lump which will collapse when the finger of the other hand is lightly pressed into it. If, when the hand is opened, the compost does not remain compressed, it must be considered that it is too dry, but on the other hand if it still remains compressed when the finger is rested on it, then it is certainly too soggy and wet.

Having now dealt with the seed-sowing receptacle, the drainage and the compost, it only remains for the actual seed to be sown. If hybridization is being done on a large scale and there are many hundreds of seeds to be sown, there may not be time to sow the seed individually, but the amateur who has probably only three or four dozen seeds could well find the time to treat every seed as an individual item, and thereby save a great deal of time at the pricking-out stage. If this method of seed sowing is considered by the grower, consideration should also be given to the individual seed pots which are on the market. These pots are made of compressed peat and have a number of advantages over the more traditional vessels used in the sowing of seed. There are two distinct types, some entirely separate and others joined together in dozens. Their most distinct advantage is that they completely eliminate the early pricking out and potting on, when very often the not-so-strong seedlings succumb to the shock of being moved. There is nothing to tell the grower that those seedlings which do fail to grow through being lost at this stage are not the most choice of all the seeds being sown. Another advantage of the small peat pots is that any irregularity in germination will not mean retaining the whole seed box or pan in the propagating frame, but only three or four small pots. The grower must remember that if peat pots are used they must be soaked well before putting the compost into them.

Before placing the seed in position the seed compost should be firmed gently. The seed is then broadcast or sown individually, whichever is more convenient to the grower, and then covered by a thin layer of the compost and firmed again. The reason for this final firming of the compost is to ensure that the seed is in close contact with it, so that the testa is softened quickly and germination is hastened.

Once the seed is sown, the whole should be given a thorough watering, and even with this simple operation there is a right and wrong way. No matter what type of receptacle is used for sowing the seed, watering should always be done from the bottom. This means that they should be left to soak in water which is deep enough to reach just below soil level, until the whole of the compost is thoroughly wet. Watering from the top, even with the finest rose on the watering can, can wash the seed deeper in the compost or leave it exposed on the surface, and either happening may well be detrimental to the success of the operation. The receptacles are then covered with a sheet of glass and a sheet of brown kraft paper, and placed in the most favourable position in the greenhouse or frame. The glass and paper will do much to prevent the soil drying out while germination is taking place. The temperature will, however, cause some condensation to take place, and for this reason the glass should be turned over daily.

It is generally accepted that fuchsia seed is erratic in germination, and if a receptacle has been used in which many seeds have been sown, the time to remove glass and kraft paper is as soon as the first seed has germinated. Pricking out the seedlings can be a critical part of the life of a fuchsia, and the grower should be in no haste to do this. The fuchsia should have at least two pairs of its true leaves. To touch a seedling at the cotyledon stage is not recommended when fuchsias are the subject. If seeds have been sown individually of course no pricking out will be necessary, and germination of the slower seeds can be hastened by keeping them covered without upsetting those seeds which have already germinated. The seedling fuchsias should be pricked out with as little disturbance to the roots as possible. Root disturbance may not mean the death of the plants, but the check it receives may delay growth for some time.

The plantlets should be pricked out into thumb pots (72's) or small peat pots, or again if they show much vigour they can be put direct into small 60's (2½-inch) pots, using J.I. Potting Compost No. 1. When watered, the new plants should be kept in a dull, close atmosphere for a few days, so that they may recover from whatever shock the move may have given them. They can then be brought into the normal greenhouse or frame surroundings, and when weather permits they may be hardened off, and grown on as normal fuchsia plants.

If pure seed from the species is used the grower will know full well what to expect, provided that he has taken the trouble to study the subject a little, and vigour should be the primary aim, with any weak or sickly plant being ruthlessly discarded. With hybrid seed there is no real way of telling what nature has intended until the plants have matured and flowered. This is one reason why every plantlet should be given some form of artificial support at an early stage of its growth, for the mixed 'bag' may produce varieties of brittle growth, 'basket' types, as well as many vigorous ones.

If greenhouse decoration only is required, the grower may decide to keep all those hybrids which flower in either their first or second season, until the time comes to replace them with choicer varieties. On the other hand, the hybridizer must be more ruthless in his selection, and only those that are considered worthy of future hybridizing or distribution to the public should be kept. All fuchsia growers will appeal to those who grow from seed not to give a name to every plant grown, thus adding further to the confusion that has existed in the fuchsia world since the plant returned to popularity.

Discarding a plant invariably means its dispatch to the bonfire, but no fuchsia should be thus dispatched without first being allowed to fight for its life. Those fuchsias thought unworthy should be planted into the border, and given a chance to prove their hardiness. It may well be that one such plant may prove so hardy as to become the parent of a new generation of fuchsias.

CUTTINGS

There are few shrubs which, when grown from cuttings, will root as easily as the fuchsia. With only a little care, there is no reason why even the most 'green' amateur should not get a 90 per cent strike, while no nurseryman would be content with less than 100 per cent. Throughout the horticultural world, when a plant or shrub is easy to grow there is always a mass of different information as to how the job should be done. This is merely because the subject being grown will grow under practically any reasonable conditions, and everybody imagines that their success is due to their special efforts, rather than the adaptability of the subject itself. The taking of fuchsia cuttings comes into this category, and there are at least six different media recommended for rooting, and to add to the confusion of the novice grower there are no two authors who agree on the merits or demerits of the use of hormone powders. To put at rest the mind of the beginner it should be said that given the right conditions, the cuttings will root with or without hormone rooting powders, and in any of the following media: sharp sand, half sand and half peat, aerated pumice stone, horticultural grade vermiculite, J.I. Seed Compost, and weathered coke breeze. What the grower has to do is to find the medium best suited to the conditions he has available, and develop his propagation accordingly. From a study of the various methods of striking fuchsia cuttings, it would appear that many of the specialist fuchsia nurserymen seem to prefer the mixture of half sand and half peat, while the use of horticultural grade vermiculite is strongly favoured by the amateur. The probable explanation as to the difference between the professional and non-professional growers is that vermiculite used on the vast scale needed by the nurserymen would prove an expensive item and would not therefore prove an economical proposition.

Much has already been written in horticultural journals concerning the use of hormone powders in assisting the rooting of cuttings. There is no doubt that these powders and solutions have done much to overcome the difficulties experienced in the horticultural world with the rooting of the shy-rooting subjects, but it does appear to be largely wasted on a subject so easy-rooting as the fuchsia. Those who advocate its use when dealing with fuchsias will quickly point out that rooting takes place more speedily when the powders are used, but this initial gain of three or four days is quickly lost when transplanting takes place, as so much root is developed by the use of these chemicals that what to the plant would normally be a slight disturbance becomes a major upheaval. The final result, however, is the same whether hormone powders have been used or not, so their use is solely at the discretion of the grower.

Propagation

There are three types of fuchsia cuttings and they are known as the softwood cutting, the semi-hardwood cutting and the hardwood cutting. The first two, the soft and semi-hardwood cuttings, are those used by practically all fuchsia growers, and the hardwood type are only used in an emergency. It is from the new spring growths that the soft cuttings are taken, and the greater mass of the annual total of young fuchsia plants are grown from this beginning. The semi-hardwood cutting is taken and rooted in the autumn from the current year's growth, which by late August or early September has developed into mature wood. This latter method is used only by those growers who have the facilities to grow the plant on during the winter months, and they are of great help to the nurserymen, who from one plant of a new variety can build up a number of stock plants from which the softer cuttings can be taken in the following spring.

A fuchsia cutting should have at least four pairs of leaves, of which the bottom pair are removed, and the cutting is trimmed with a razor blade, or a very sharp knife, just below the joint from which the pair of leaves are removed. Some growers do, in the case of hardwood and semi-hardwood cuttings, prefer to cut or pull the cuttings from the parent plant, so that a little piece of the main stem comes away with it. This small attachment is known in horticultural circles as a 'heel', and this is trimmed up so that only a thin part of the cambium layer is left attached to the cutting. The growers who prefer to do this feel that this type of cutting is quicker rooting when the wood is a little on the hard side, but it is true to say that fuchsias will root easily with or without the heel.

As in seed sowing, the receptacle in which to root the cuttings is very much at the discretion of the grower, who will have to take into consideration the number of cuttings he intends to root. The nurseryman with his 'Mist Propagation', which is described later in this chapter, will use the greenhouse staging itself, while the amateur will find the small 60 pots more convenient. Seed boxes, seed pans, and a host of other utensils can be used according to circumstances.

The amateur will find the 3-inch pot (small 60) the most convenient for general use for it will take five cuttings quite comfortably, and if one pot is used for one variety, there is less chance of the cuttings and labels becoming mixed. There is nothing more frustrating than growing on a plant in the belief that it is a certain variety, only to find on flowering that it is something entirely different.

The cuttings having been trimmed up, and their lower leaves removed, they are placed round the edge of the pot and firmed into the rooting medium. Do not press the cuttings into the medium as the softwood cuttings will often be damaged, and the hardwood ones may well fall out when water is applied. A pencil-thin dibber, or a pencil itself, should be used to make a hole for the cutting, and it should be placed into the hole so that the lowest of the remaining leaves are just above the surface of the medium. As soon as the cuttings are firmed, the whole should be given a good watering prior to being put into the propagating frame. The water should be applied gently so as not to disturb the cuttings.

The length of time taken by fuchsia cuttings to form roots varies according to the temperature and the general conditions into which the cuttings are placed. Put into a propagating frame with a bottom heat, best obtained by using one of the many electric soil-warming units now being produced so cheaply for the amateur gardener, an air temperature of 60° F., and plenty of atmospheric moisture, roots will have formed in approximately nine to ten days. On the other hand, fuchsias will root easily in early summer in a closed cold frame, but rooting in this case can take anything from fourteen to twenty-one days.

There are certain rules which must be obeyed to ensure that the whole operation is one hundred per cent successful. The cuttings must be shaded from strong sunlight, and at no time must they be allowed to become dry either at the rooting medium or in the air surrounding the remaining leaves. Any leaves which become detached from the cuttings must be removed and as soon as rooting has taken place the new plants should be given air.

If the media being used have no natural sources of nutrients in them, i.e. sand or vermiculite, then potting on into J.I. Potting Compost No. 1 should be done as soon as a reasonable amount of root has been grown. Leaving too long in such media will result in a weak root system, and a spindly top growth, and the chances of getting worthwhile results from such plants are slim indeed.

Mist Propagation

The modern nurseryman has to produce many hundreds of young plants for the spring and early summer market every year, and this mass production of rooting cuttings is done quickly and economically by what is called 'mist propagation'. For this purpose he usually has one greenhouse devoted entirely to the propagation of his plants. An example of the layout of such a house would mean that the centre staging would be of such construction as to hold many hundreds of cuttings either in boxes or dibbled into the rooting medium direct on to the staging. The stagings along the sides of the house would be there for the purpose of receiving the rooted cuttings that have been potted on until they have taken a firm hold of their new compost prior to moving them to a cooler greenhouse or the frame to harden off.

The actual mechanisms and principles involved in 'mist propagation' are complex, and its erection needs some engineering skill. Briefly and without intricate technicalities, it consists of a water supply pumped up a number of vertical pipes; at the top of each of them is an extremely fine mist spray. Also there is a surface of metal which is called an 'artificial leaf', above which is an electrical contact. When the water supply is turned on the fine mist from the spray covers a certain area with moisture, including the 'artificial leaf'. The amount of water on the 'leaf' builds up all the time the mist is being emitted from the spray, until it meets the contact. The effect of the water touching the electrical contact immediately breaks the circuit and switches the pumps off, thereby stopping the supply of any further water. The water remains switched off until evaporation has once again freed the electrical point from its contact with the water, and the circuit is restored, and the spray is once more pumped out. This procedure goes on indefinitely.

It can be understood that the evaporation from the 'artificial leaf' will be identical with that of the leaves of the fuchsia cuttings being propagated. This means that the respiration rate of the cuttings is nil and their supply of life-giving air is at the maximum. The whole complex system of life that Nature has given to living plants can now, in the case of the fuchsia cutting, be concentrated on the production of roots.

The rooting medium for use with 'mist propagation' should be of a neutral nature so that it can be used for several successive batches of cuttings. This means that loam should not be used. A mixture of sand and peat to which is added a little charcoal to maintain sweetness is probably the most satisfactory medium to use.

All rooted cuttings should be hardened off as soon as possible after rooting. Kept in the atmosphere that was used for propagating itself, the fuchsia will tend to grow too much top growth at the expense of root development.

Having mentioned the mass production of rooting cuttings by 'mist propagation', it might be useful to the smaller grower who only wants two or three cuttings from one or two plants, and who lacks the facilities of a garden, to mention the use of polythene bags. The Fuchsia Society boasts of the vast number of window-sill gardeners among its members, but there must be many thousands of others who grow their fuchsias in the same way and who feel that the taking of cuttings is only for those with a greenhouse and frame. Fuchsia cuttings will root quickly and easily in a 3-inch pot of vermiculite placed in a polythene bag, its ends sealed with an elastic band. The whole can be placed on a window sill, but not in direct sunlight, in a warm room, and the only attention apart from the initial watering is the turning inside out of the bag every three or four days. This action is merely to give the plant a complete change of air and to prevent the leaves rotting by coming in contact with the condensation on the polythene.

This chapter has tried to show that every grower of the fuchsia has the facilities to propagate the plant, and if it is remembered that under restricted conditions the young plants are the best plants, there is no reason why the fuchsia should not always be seen at its best.

CHAPTER IX

Exhibiting

Many people of this age will doubt the potential of the fuchsia as an exhibition flower, but anybody who has seen the shows of the British Fuchsia Society, or the exhibitions displayed by fuchsia nurserymen at the Royal Horticultural Society's Halls and the Chelsea Flower Show, will know what a tremendous attraction they can be.

The golden age of fuchsia exhibitions was between 1880 and 1910, when huge pyramids fully 10 feet high were exhibited first by that great plantsman and raiser of new fuchsias, James Lye, and later by his son-in-law G. Bright, raiser of that outstanding hybrid 'Pink Pearl'. After the First World War, the few fuchsias that were exhibited had little to commend them and the idea was created that the fuchsia was not an exhibition plant.

The formation of the British Fuchsia Society in 1938 brought a new impetus to exhibiting, and the stands of outdoor-grown fuchsias displayed by the late Mr. W. P. Wood up to the time of his death in 1955 did much to show gardeners how plants of exhibition quality could be grown even without the aid of a greenhouse. Many more displays were put up by specialist nurserymen, culminating in a number of 'Gold Medal' displays at Chelsea Flower Shows by the firm of L. R. Russell of Windlesham. These displays and the show of the B.F.S. had once again established the fuchsia as a first-rate subject for exhibition.

Exhibiting of anything, whether it be fuchsias or dogs, has a great deal to do with the popularity of the subject being shown. Good displays of a flower will attract attention, and because of this attention more people will want to grow the flower. Members of the committee of the B.F.S., nurserymen, and everybody who has the furtherance of the fuchsia at heart, realize all this and besides exhibiting themselves they provide the stimulus for others to grow.

Fuchsia growers are fortunate inasmuch that all varieties of fuchsia, provided they are well grown, are suitable for exhibiting. There are a number of varieties which are more suitable than others, because they are either more showy in appearance, or more floriferous, or have a more pleasing habit of growth. These are listed in Chapter XXI at the end of this book.

It should be the aim of every fuchsia enthusiast to show his collection to as many people as possible as often as possible. Well-grown fuchsias are in themselves a work of art, and like the conventional arts they should be continually displayed in order to bring enjoyment to everybody who would appreciate them. This means that the plants should either be brought to those who want to see them, or the people be brought to the plants. There are many growers who enjoy people visiting them to see their collection of fuchsias, but the plants are always seen at their best, and by far more people, when displayed in their full glory on the exhibition table. An exhibition plant should be only the best. It must always be remembered that a poor plant will bring no credit to either the flower or the person who exhibits it.

The first essential on the road to a 'First Prize' exhibit is good cultivation. The plants should be grown as exhibition plants from the time they are started into growth in the spring until the date of the show. They must receive no check in their growth from the time they are re-potted into their flowering-size pot.

Fuchsias can be exhibited in any size pot, but the majority of the Show schedules stipulate that the pot size should not exceed 6¼ inches. This is done with the purpose of giving everybody who wishes to exhibit an equal chance. A 6¼-inch pot is a fairly easy size plant to transport to the show as well as showing the flowers off to the best advantage. In these days of the smaller greenhouse and garden there are indeed few people who could find room to house the grand plants of yesteryear.

The cultivation of fuchsias has already been dealt with fully in this book and the cultivation of exhibition plants is no different, except that it is done very thoroughly. The training of the plants also calls for greater care and attention. The effect of casually looping all the growths to a stick pushed into the centre of the pot may well be camouflaged in a packed greenhouse, but on the show bench it will stick out like a sore thumb.

Bush plants are the most common form of fuchsia exhibited, and are, unfortunately, usually the worst trained. The majority of bush-trained plants, if they are well grown, will require four or five canes to support their growth correctly, for apart from any other eventuality they are to travel to the show heavy with flower. These canes should be put into the pot from the time the fuchsia is potted into its flowering-size pot. If they are pushed into the compost round the edge of the pot at that time it will prevent damage being sustained by the roots when later on the plants are pot-bound. Generally the canes are placed four at equal distance from each other sloping at the same angle as the flowerpot does from its top to its bottom, and one cane standing upright in the centre of the plant. Good-quality bass is extremely strong and it should be used as thin as is practical so that it is obscured by the plant as it grows. The practice should be to tie with as little as possible as often as is necessary. Too often we see a plant which has just a few heavy ties to support it, thereby ruining the whole effect of daintiness which is the fuchsia's charm.

The whole operation is now simply this. A strong central growth is chosen for the centre cane and four growths of reasonable strength are chosen to grow supported by the four canes around the edge of the pot. All other growths that develop are tied inconspicuously to either the already supported growths or the canes, whichever is nearest. As the plant develops and, through pinching out the growing tips, more branches begin to grow, support those branches by tying with bass to the nearest supported growth. To keep returning to the canes for support may mean dozens of long lengths of bass, which would be difficult if not impossible to hide, whereas an already supported branch may be only a fraction of an inch or so away, and the tie to this could easily, on the show bench, be covered by a leaf. The grower must keep one growth growing up the supporting canes, and when he pinches out these growths one of the new shoots should be selected to continue up the cane. Kept close to the cane, the branch will practically obscure it. It is important that the canes chosen are longer than the anticipated height of the fuchsia when it is fully grown, so that the whole operation is not ruined by having to re-tie the complete plant before the show date. When the plant is arranged neatly on the show table it is a simple thing to take the secateurs and cut the canes to approximately three inches below the leading growth.

Many may wonder how a good exhibition plant is recognized. The exhibitor must remember that he is presenting a flower to the judges and the public and therefore the flowers on the plant must be as prolific as the variety will allow and they must be of good quality. The growth and foliage are there to present and support the flowers and they must be without blemish so as not to distract the judges' eyes from the flowers being shown. What the exhibitor is actually doing with each plant is presenting a symmetrical floral display complete with supporting foliage.

Timing fuchsias for exhibition has always been a problem to those growers who have not studied the matter thoroughly. To get a plant of good shape and to get the maximum growth supporting the maximum amount of flower, it is often necessary to pinch out the stronger growths of a plant several times. All authorities on fuchsias admit that in normal summer weather it takes approximately six weeks from the time the tip is pinched out until the resultant shoots flower. This statement has been made in many publications in the past and while it is very true in its fact, it is far too vague for the beginner to understand. What it means is that the shoot that is stopped already has the green flower buds on it and these are removed with the pinch. Six weeks later the shoots that come as a result of this stopping will be in flower, but only just so. A shoot that has no sign of bud on it whatsoever should not be stopped, unless it is necessary to maintain the shape of the plant, for to the six-week period must be added the period that would have been taken to produce bud on the original shoot.

When buds first show on one shoot it is practically certain that within a week or ten days buds will appear on all the growing tips, such is the hormone activity within the plant. Therefore the stronger-growing tips should all be stopped, provided that one or two shoots are showing bud at least eight weeks before the show. This period not only gives time for the new growths to make flower but also the secondary and supporting growths. It must appear obvious that if the first blooms come in six weeks the main flush will follow in approximately a fortnight hence. There must be very few people who have not seen the same thing happen on the floribunda rose, which starts flowering intermittently before bursting into a blaze of colour.

Good judges of fuchsias are, unfortunately, rare. The judges chosen by the British Fuchsia Society to judge their main and subsidiary shows are household names among fuchsia enthusiasts, and their judgement is rarely, if ever, in doubt. At local shows it is impossible to get specialists in every section of gardening, and many a good all-round judge has through ignorance fallen into the error of going by size of plant and bloom, rather than quality and quantity. However, as more and more fuchsias are exhibited, so the standard must improve.

Fuchsia judges of experience will look well for the following points. First, the quality of the plant generally, and then the quality of the blooms and the floriferousness according to the variety. A variety such as 'Snowcap' they will expect to have many dozens of blooms of good quality, while one such as 'So Big' may have only one dozen but of good size, and again of good quality. The foliage must be of good colour, again according to the variety, and any yellow or bug-eaten leaves must be removed before the judging takes place. Symmetry of shape is looked for, and all ties and supports should be invisible at first glance. Should the schedule call for more than the one plant in a single exhibit the plants should be uniform in size and shape. If the schedule calls for three or more plants the exhibitor can often overcome a lack of uniformity by displaying the plants at different levels. For example, if one plant is taller than the other two in a three-plant exhibit, put the two shorter plants side by side in front and raise the tall one well above their level by putting it on an inverted flowerpot or a neat block of wood. This will often give the appearance that the centre plant is only tall because of its pedestal. Naming the varieties in the exhibits is important in the bigger shows. Judges realize that visitors to the show love to know exactly what they are looking at. Discussions on varieties and their merits help to create a show, and it is often said that the success of a flower show can be judged by the number of visitors who can be seen scribbling down names in their notebooks or diaries. If the name of a certain variety of fuchsia is not known by an exhibitor, it is flattering to the judges to see a card marked 'JUDGES—PLEASE NAME'. Admittedly they may not know the variety themselves, but they will often endeavour to find out if any of the knowledgeable people, always present on such occasions, can help.

The greatest drawback to the fuchsia as an exhibition plant is the difficulty in transporting it to the show. Exhibition plants, even when restricted to a 6¼-inch pot, can reach a considerable size, and two is the utmost that can be carried by the largest family car. With local shows the difficulty can be overcome by making frequent trips to and from the Hall, but when the show is, for example, at the R.H.S. Halls in London, the keen exhibitor must make other plans. When taking twenty or thirty plants any distance the ideal means of conveying them is by the conventional furniture van. This form of transport is roomy, is low to the ground for easy loading and unloading, can travel partly open at the rear to allow the plants to have some air, and above all is softly sprung, which means that the plant is not shaken to pieces before the hall is reached. The cost of transport may well be prohibitive to one person but it is a fact that where there is one fuchsia enthusiast, within a very short time there are two, and then three. Two exhibitors sharing the cost of transport will bring the cost down to a reasonable level.

The packing of the plants in the transport may, to many people, present some form of problem. In the past keen exhibitors have thought up the most intricate methods of wrapping up or supporting their plants. Throughout life we are continually reminded that very often the simplest way of doing things is the best way, and in the transporting of exhibition fuchsias this rule definitely applies. When there is a large plant in a small pot the problem is one of top-heaviness, so some artificial way of extending the base must be devised to counteract this.

A keen grower of fuchsias in the Selsdon area of Surrey devised a simple frame, details of which are illustrated overleaf, which enabled the grower to carry two plants at once with ease. As can be seen by the diagram, this is an extremely simple piece of equipment for anyone to construct, and when the plants are packed so that the plants in every frame are touching each other to give further support, many plants can safely be

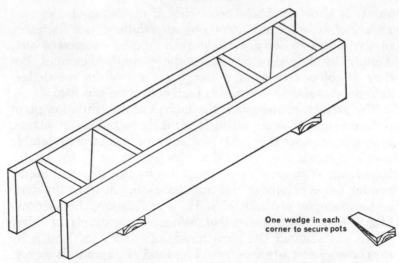

One wedge in each corner to secure pots

Device For Carrying Two Fuchsias To Exhibitions

transported any distance. Many trips to the R.H.S. Halls at Westminster have proved the effectiveness of this method.

Within the last few years with the growing enthusiasm being shown by the smaller exhibitors, and the increasing number of shows at which fuchsias can be exhibited, a method of transporting a small number of plants has had to be devised whereby the grower can use his own car. It has been found that by wrapping each plant individually with either butter muslin, builders' cleaning cloths, or tissue paper, and then packing the plants tightly into the vehicle so that all the plants are supported by each other, it is possible to carry as many as a dozen good-size exhibition plants a considerable distance without damage or loss of flower. The huge baskets and standard plants, of course, do not lend themselves to being squeezed into the normal family car, but the bush plants, even those of some substance, are being carried effectively by this method.

Anybody wishing to exhibit fuchsias successfully can only expect to be successful if the utmost attention is given to detail. To grow an outstanding plant may be expected of any grower, but in addition to this an exhibitor is expected to give a first-class presentation of that plant. The flowerpots in which the plants are exhibited should be clean and of good colour. Some clay pots as they get older become more and more difficult to clean owing to the surface becoming terribly pitted. To overcome this the grower should get from his local paint merchant a pound of dry red ochre and a pint of petrifying liquid, the cost of both items together being no more than five shillings, and make a paste by mixing the red ochre with the petrifying liquid. If this paste is rubbed or brushed over the outside of the difficult pots it will obliterate all unsightly marks, and make the pot look new again. There is no fear of the colour coming off with watering or handling as it will take some time to break down, by which time the season will be over.

A well-grown bush-shaped fuchsia will completely hide the compost in which it is grown, but there are some very upright in their habit and these tend to show more of the pot and the compost in it. It is these varieties and the standard plants which require some special attention prior to judging. After the plants have been staged and given their final watering, a sprinkling of fresh compost should be given over the old compost, so as to freshen up the appearance of the plants. Some exhibitors use peat and others prefer moss, but it matters little which actual material is used if the old, and very often sour, soil is covered.

Many a first-class fuchsia has been disqualified through the failure of the grower to study and understand the schedule. If there is anything in the schedule that is not fully understood that point should be queried with the show officers. To take a chance may only mean unnecessary disappointment and useless waste of effort. In exhibiting fuchsias special note should be taken of the size of flowerpot permitted in each class, to ensure that this is not exceeded. Some difficulty is experienced by novice exhibitors in the showing of semi-double varieties. In some seasons some of the varieties throw flowers which are very near to single varieties in their appearance, and if the classification of the variety is not known there is some doubt as to what classes they should be exhibited in, singles or doubles. If the grower is uncertain concerning the classification of any variety he should endeavour to find out the facts as early in the season as possible so that he can work out the classes he is to enter and plan accordingly. All semi-double varieties are classed as double for the purpose of showing.

If a basket is being shown then the exhibitor should take

special note of the size of the basket permitted and also the number of plants that are allowed to each basket. Standard plants, table, half and full, are measured from where the stem enters the compost to the first branch of the head on the stem. The amount of clear stem should be as follows: table standards, 10–17 inches, half standards, 18–29 inches, and the full standard should not be less than 30 inches, and should not exceed 42 inches.

The plants will have been well watered prior to their being transported to the show, but give them a little more water after staging, for halls and marquees can become unbearably hot, and under such conditions the fuchsia's rate of respiration is rapid and in consequence the plant could well begin to flag before the judge has seen it. Seed-pods of dead flowers will in the main have been removed well before the show, but it is as well to go over the plant after it has been staged in case any have been missed.

After having staged the exhibits, the exhibitor should stand back to ensure that the best side of the fuchsias is facing the front. This probably will not fool the judges, but it does give the best effect to the exhibits and the paying public deserve only the best.

The British Fuchsia Society, realizing that the main shows where fuchsias are shown are not always within easy reach of all growers, have offered a 'Blue Ribbon' to any Horticultural Society that includes within its show schedule a class for fuchsias. Furthermore they are promoting provincial shows, in conjunction with leading Horticultural Societies, and trophies and medals presented by them are eagerly contested.

Exhibiting fuchsias is fun. Nobody has yet made a fortune either out of arranging the shows or exhibiting in them. The exhibitors are always financially out of pocket after paying for transport and fares, and the officers of the show and the judges give a great deal of very often valuable time for no monetary reward whatsoever. As nobody is in this business for what he can get out of it, but only what he can give to it, it is essential that whatever decisions are reached should be accepted in the correct spirit. Appeals against judges' decisions should only be made after discussion with officials of the show has failed to reveal the reasons for such a decision. Damaged pride is the only cause for bad feeling at any flower show and the quickest way to repair it is to congratulate the person who took the prize.

CHAPTER X

Hybridization

The first record of the hybridization of the fuchsia took place in 1825, when it is recorded that *F. coccinea* and *F. magellanica* were crossed with *F. arborescens*. What history has failed to record is the results achieved by these crosses. However, this was the beginning of a great flood of fuchsia hybrids which was to go on up to the early twentieth century, and then after a lapse of several years, when the popularity of the flower waned, to return to a flood soon after the end of the Second World War.

The first hybrids were the work of the British nurserymen, who were quickly followed by the French, and then the Germans. With the return of the fuchsia to popularity after the Second World War it was the Americans who took the initiative in this field, and it is to their credit that the majority of the worthwhile introductions of recent years are the work of some of the great American plantsmen.

Chapter XI gives the list of species that have so far been used in the raising of hybrids as far as is known. It is indeed unfortunate that the early raisers of fuchsias either kept no records of their crosses, or if they did their records have never been found, and this may well mean that some varieties have the blood of some species in them of which the modern hybridizer is not aware.

It can be said that once a grower has mastered the art of growing fuchsias, his thoughts will often turn to producing varieties of his own. Many will say, perish the thought, and leave it at that, others will have a half-hearted try and at the first sign of failure give up, but the true enthusiast will be bitten by this fascinating 'bug', and after a few hit-and-miss attempts will get down to the job with a real purpose.

Before suggesting the purposes that might be pursued it would perhaps assist if some of the principles of hybridizing were explained and understood. Although it is obvious by the results that the early hybridizers of note such as Lye, Bland and Story had certain objectives in mind when they produced their popular varieties, up to the time of the discovery of the theory advanced by Mendel all hybridizing was hit-and-miss in character.

Gregor Johann Mendel (1822–84) was an Austrian monk, devoted to his calling, and like many of those in holy orders in that day a specialist in other fields. He was a student of botany and mathematics, both subjects which were to assist him tremendously in evolving the theory that was to revolutionize not only the hybridizing of plants, but also many other subjects. In his early experiments Mendel chose as his 'guide stick' the culinary pea, and briefly his findings were these. Having crossed a tall yellow-seeded strain of pea with a dwarf green-seeded one, he found that the offspring were all tall, yellow-seeded. These features he called 'dominant', and the plants themselves were known as F_1 hybrids, which means the hybrids of the first filial generation. Crossing the resultant seedlings with themselves showed results of the second filial generation, F_2, as 75 per cent tall and yellow-seeded, and 25 per cent dwarf and green-seeded. He therefore called the features that had disappeared at F_1 and reappeared at F_2, that is the dwarfing and green-seeding, 'recessive' characters. The result of the F_3 generation showed that the seeds of F_2 contained 25 per cent tall and fixed, 25 per cent dwarf and fixed, and 50 per cent unfixed, the latter again breaking up into those same percentages after F_3. Throughout his experiments Mendel established that when these crosses were studied for several generations the results followed some quite elementary mathematical laws.

All hybridizers, whether they be working with fuchsia or something else, have cause to be thankful to Gregor Mendel for what are now known as the 'Mendelian Laws' which were first put forward in 1867; and to Professor William Bateson, the man responsible for giving them worldwide recognition in 1900.

In 1955 a team of scientists from Manchester University under the direction of Professor Harland and Mr. S. K. Chaudhuri M.Sc. (Calcutta) applied Mendel's theory to the fuchsia, and it is interesting as to the results achieved. By way of experiment they crossed a species *F. lycioides*, which is a small flower of hardy species with the traditional fuchsia colours of red and purple, with that variant of *F. magellanica* known as *F. magellanica* var. *molinae*, commonly called '*mag. alba*', which has an almost white tube and sepals and a very light mauve corolla. Furthermore, the current season's stems on *F. lycioides* are red in colour while those on '*mag. alba*' are green. The F1 hybrids from this cross were all of *F. lycioides* colour in both flower and stems. The next cross, F1 × '*mag. alba*', showed clear segregation into 12 plants in which the red colour of the stems was present and 13 plants from which it was completely absent.

Once the Mendelian Laws were understood by the scientists there followed a great research into the genetics and cytology of plant life so that these laws could be explained, and this led very quickly to the discovery and understanding of chromosomes and genes. A detailed explanation of these things would make this chapter look like a large algebra problem, but it is essential that a hybridizer who wishes to know exactly what he is about should know the elementary rules involved. Cytology is the study of the cell structure of plant life and involves the study of chromosomes and genes.

Nature invariably works to a pattern, and to certain laws. And to guard these laws and to preserve the balance of Nature, it has chromosome values and incompatibility, so that only kind can breed with kind and the breeding of monsters is prevented. Chromosomes can be crudely described as microscopic strings within each plant cell along each of which are the genes. Science tells us that these genes are responsible for the characteristics of all forms of life. By bringing about various chemical changes they dictate the colour, size, shape, in fact everything that goes to make a plant. The genes, we as hybridizers can do nothing about, but as they are in fact part of the chromosome make-up, and since chromosome values can be altered, it is to them that we must give attention.

The species of fuchsia have in the main 2 sets of 11 chromosomes ($2 \times = 22$) and because they have 2 sets are known as 'Diploids'. *F. magellanica* and its variants and *F. lycioides* are exceptions, having 4 sets of 11 chromosomes ($4 \times = 44$) and they are known as 'Tetraploids'. This means that every cell in

F. splendens, for example, has 11 pairs of chromosomes while every cell in *F. lycioides* has 22 pairs. However, when the sex cells, or gametes, are formed—and they are the pollen on the anthers and the ovules in the ovary—a different kind of cell division takes place known as meiosis, and instead of every chromosome being divided in two, the pairs divide so that each sex cell has just one set of chromosomes instead of a pair. Nature's reason for this is quite simple to understand, for when the male cell with its one set joins with the female cell which also has one set, the cells which are to form the new seedling will once again contain a pair of sets. To take a very simple example of hybridizing, if a cross is made between *F. magellanica*, the ovule of which would have 22 chromosomes, and *F. fulgens*, the pollen of which has 11 chromosomes, the resultant seedling would have 33 chromosomes and would therefore be known as a 'Triploid'. The names given to the grade of chromosome duplication are as follows:

Diploid	(2 sets of chromosomes)
Triploid	(3 sets of chromosomes)
Tetraploid	(4 sets of chromosomes)
Pentaploid	(5 sets of chromosomes)
Hexaploid	(6 sets of chromosomes)
Heptaploid	(7 sets of chromosomes)
Octoploid	(8 sets of chromosomes)

All this knowledge may seem unimportant to the individual who just wishes to cross two varieties to see what they will produce, but even that person will in time be wishing to go into the matter from a more scientific angle, so that he or she can choose the name for what one day may be a famous variety.

Basically the plants with the even number are those most fertile, while those with the odd numbers $3\times$, $5\times$ and $7\times$, are usually sterile; the reason being, of course, that the chromosomes cannot divide if the numbers are odd. The raiser of new fuchsias is, in this respect, much luckier than raisers of many other plants and shrubs, because there are a limited number of triploids, pentaploids and heptaploids among fuchsias, which do naturally what professional raisers do with the aid of colchicine, that is, double the chromosome value of the sex cells. Many of the F1 hybrids of *F. magellanica* and *F. fulgens* do this repeatedly, and because of this there are a whole host of beautiful hybrids that can trace their ancestry back to the cross between these two species. Those near-white varieties 'Rolla' ($7\times$) and 'Countess of Aberdeen' ($5\times$) are continually being used as parents by hybridizers trying to find that perfect all-white variety, and their progeny, although having perhaps fallen short of the perfect white, number among them many worthwhile and well-known varieties.

To sum up on chromosomes it can be said that the diploids are the majority of fuchsia species, the triploids the results of crosses between diploids and tetraploids, the tetraploids some of the hardier species and the first of the well-known hybrids which can be called the 'bread and butter' varieties, such as 'Marinka', 'Loveliness' and 'Lustre'. At the top of the table there are the octoploids ($8\times$) and this is the group which contains those big double varieties so admired by all those who see fuchsia collections. Among these are the varieties 'San Leandro', 'Pacific Grove' and 'Purple Heart'.

Before anybody can start to think of hybridizing there must be an understanding as to the composition of the flower with which we are to deal and the functions of its various parts. First there is the pedicle, which one would normally call the 'stalk'. This is either red or green in colour according to the variety or species, and supports the flowers, besides being the channel through which pass the nutrients for its well-being. This joins the flower proper at the ovary, or seed vessel; this again can be red or green and contained within it are the ovules, which eventually become the seeds that are to be sown. From the ovary extends the 'tube', which on any other flower would be known as the base of the calyx. This can be any colour except, unfortunately, blue. The tube is merely a protection to the long stamens and pistil, and it leads down to the four sepals which collectively can be called the calyx. Botanically speaking these are modified leaves whose purpose is to protect the delicate petals, stigma and anthers until they are ready to burst forth for the purpose of fertilization, but on the fuchsia they are also an important part of the decorative scheme, as is the tube. Grouped beneath the sepals is the corolla. This is the collective name for a group of petals, and on single varieties there are four petals in a circle, and in the doubles there can be any number above four. The petals of the corolla contain the most delicate and richest of colouring of which the fuchsia can boast. Red, blue, orange, purple and any amount of combinations of these colours, but still no yellow. The corolla surrounds the sex organs of the flower, the stamens and the pistil, and like all the functional parts of a fuchsia they too play an important part in the appearance of the fully matured flower. The stamens, usually eight in number, consist of thin 'stalks' on the end of which are the anthers, pads which emit the pollen. They in turn surround the pistil which has a much thicker stalk known as the style, and on the end of which is a sticky protuberance, the stigma.

Nature has extended the pistil beyond the stamens in the fuchsia for a definite purpose. As the flowers swing in the wind, or are rocked by an alighting insect, the pollen falls and in falling has to pass the tacky stigma to which it sticks, and thereby fertilizes the seeds.

Thought of purpose should always be the first stage in hybridizing. To hybridize without an object in mind is similar to filling in a football coupon; luck may prevail, although it is very unlikely, and there is no satisfaction like that of an objective achieved. Having fixed a target at which to aim the next thought is how best to achieve the desired result. This is really the hardest part of the whole operation, for it often means hunting through catalogues or other people's collections to find a species or variety you require. The parents selected for crossing should be started into growth as early in the growing year as possible. No attempt should be made at training or stopping the plants, as the aim should be to produce flowers as early as is feasible. Do not pot the plants beyond the 5-inch pot stage, and give only enough feed to maintain them in a healthy condition. The sooner the plants become pot-bound the sooner they produce flowers, and excess fertilizer will only help to produce growth useless for the purpose in hand. Too much heat will only produce a lot of unwanted growth, so grow the plants as cool as possible. Plants started into growth by gentle heat in early March can often be made to flower by early May, which means that with all things going well new seedlings can be raised to a reasonable level the first year.

As the first flowers open on both parent fuchsias they should be left untouched, as these are the blooms which are needed to

supply the pollen. The pollen on a fuchsia takes some little time
to mature and it probably would not be ready for use until the
third or fourth flowers on each plant are starting to open. As
soon as it is seen that the pollen is ripe—this can be seen by a
fine powder falling from the anthers, or can be tested by rubbing
a fingernail over the anther so that some pollen is visible on
the fingernail—select a bud on each parent which is on the
point of bursting open. Assist the opening of the buds by 'pop-
ping' them between the fingers and gently ease back the sepals,
revealing the corolla, anthers and stigma. With a pointed pair
of sharp scissors carefully remove the anthers by cutting the
filaments, leaving only the style and stigma surrounded by the
corolla. Do not fertilize immediately, but put a small fine-mesh
muslin bag over the blooms from which the anthers have been
removed. The stigma is not at its tackiest at the initial opening
of the bud, and two days' development will make it much more
receptive to the pollen. Some authorities suggest the use of a
polythene bag in place of the fine-mesh muslin bag, but condensation, for which polythene is notorious, can quickly rot the
style and ruin the whole operation.

For the actual act of fertilization, remove from one of the
plants a flower with the ripe pollen and rest it on a plate beneath
the fuchsia which is to bear the seed. Now remove the muslin
bag from the seed parent and immediately pick up the flower
with the ripe pollen and brush the anthers lightly over the
stigma until it is covered in pollen. Replace the muslin bag
immediately, and throw away the flower used as the pollen
agent as it is of no further use. This same operation should now
be done again, but the fuchsia plant which acted as the male
in the first instance should now be the female and that which
was the female should now be the male. Simply A should be
crossed with B and B should be crossed with A.

It is essential at this stage that a label be put on each muslin
bag giving details of a crossing and the date that it was made. It
is very wrong to even try to memorize details such as this, and
to do the job properly a book should be kept as well giving
every detail possible so that the raiser can continually refer
back to whatever the original parentage was.

Assuming that species or varieties used in this operation were
compatible with each other, within a short time of the pollen

being transferred to the stigma, one of Nature's greatest wonders will take place. Under stimulation from the stigma the pollen grains 'germinate' and grow a microscopic tube which pushes its way down the style into the ovary and then into the ovum where the female nucleus lies. When the tube is complete the male gamete will pass through the tube and join the female. So a fertile seed comes into being.

From the time of fertilization the flower within the muslin bag will begin to fade and it will eventually drop off, leaving the swelling seed berries still attached to the plant. For the sake of hygiene remove the dead flowers from the bags but replace the bags so that the berries can ripen therein. This action makes no difference to the seeds or the berry, but should the berry leave the plant when the raiser is not present, the muslin bag, and the identifying marks on it, will make it easier for him to find and identify.

It can be assumed that a berry that had become detached from a plant of its own accord contains ripened seed, so the operation of extracting the seed should now be performed. The fuchsia berry contains four compartments, and these can be seen quite easily on most ripened berries. With a razor blade carefully cut the berry into four, making the cut between the bulges that indicate the separate compartments. Each portion should now contain seed, and the quarter should be crushed by pressing it with the thumb on to a piece of blotting paper. This is a much more satisfactory way of extracting seed than that advocated by many, crushing the whole seed in one operation. Some fuchsias produce enormous squashy berries which may only contain two or three seeds, and these are often difficult to find among the mass of debris that was once a berry.

A good magnifying glass will soon help the raiser to sort out the seed from useless husks. The amount of seed to expect from a cross depends a great deal on the species or varieties used. Some contain a dozen or more, while others may only produce two or three. The size of the berry is no indication of what to expect inside, for often it has been shown that a large berry will contain less than a much smaller one of a different variety.

Once the seed is extracted from the berry it should be placed on blotting paper and placed in an airy place to dry. The sowing of the seed should be undertaken at once if this is at all possible, although if the harvesting takes place after the month of August it would be better to keep the seed over the winter months in an airtight tin, and undertake the sowing as early in the spring as possible.

There has always been a doubt concerning the time that the fuchsia seed remains viable, that is, capable of germination and growth. Generally speaking the seed loses much of its viability rather quickly and therefore the sooner it is sown the greater chance there is of it germinating and growing. Those who have purchased fuchsia seed from commercial growers will have noticed how uncertain germination is, although success varies from packet to packet. Seed harvested and sown in August is practically certain to succeed, and if a temperature of 55°F. can be maintained throughout the winter it is reasonable to expect the plants to produce their flowers by the following spring.

To produce something worthwhile at the first cross is most unlikely, and sometimes the characteristics that the breeder himself is trying to introduce into his fuchsias may seem even less apparent than when he started. This does not mean that the resultant seedlings should be thrown on to the scrap-heap, for as Gregor Mendel proved to the world, there are many features of a plant which are 'recessive' in character, and these characters may well appear in the next generation.

The next step is to cross the seedlings with each other in order to bring out the features the grower is aiming at. The second generation may well produce a fuchsia worthy of the name, and it is at this stage that a 'weed-out' of the seedlings would be advantageous. Weak growers, those with inconspicuous flowers, and anything which the grower feels is not at least as good as the original parents should be discarded. Only those plants which can be termed worthwhile varieties or which have a feature in them that the hybridizer feels would be useful for further breeding should be retained. The grower should not make the mistake of giving a name to everything he raises until he is quite sure that there is not an identical fuchsia already on the market, or that he has not raised an inferior copy of an existing type. In this respect he should seek the aid of the British Fuchsia Society, who are in a position to sort this matter out for him.

Hybridization

What should the fuchsia hybridizer aim for when breeding new fuchsias? There will never be a perfect fuchsia as there will never be a perfect rose or chrysanthemum, simply because two people's ideas of perfection will never be the same, so here the scope is unlimited. There are, however, many features which all fuchsia lovers would like to see in their plants, and the hybridizer would be doing a service to them all if only one of the objects was achieved.

Of top priority must surely be the introduction of a large-flowered fuchsia with the hardiness of F. Riccartonii. Many of the large-flowered varieties have *F. magellanica* in their family tree as has F. Riccartonii, and somehow this hardiness has got to become the dominant feature. Many fuchsia growers feel that if the time came when the pot-grown fuchsia of the local market were of a type such that, while retaining its flower size, it could be planted into the back garden after its spell in the house, and therefore become a permanent feature of the garden, then its popularity would exceed that of any other flower with the possible exception of the rose. This is no dream, as many of the large-flowered hybrids do already border on the edge of hardiness. It has often been recorded that when there has been a run of several milder-type winters, some of the so-called 'tender' varieties of fuchsia have lived and flourished in the open without any protection whatsoever.

Hybridizers from earliest times have always set out to breed something which on the surface has appeared impossible. Targets such as a pure blue rose or scarlet delphinium have fascinated man for generations, and one has only to see the latest introductions in these fields to realize how near he has got to that target. The fuchsia hybridizer in the past has also aimed at what to many was the impossible, such as the pure white fuchsia and the introduction of yellow into either the sepals or corolla or even both. The pure white, largely due to the American hybridizers, is now almost a reality. The amount of colouring other than white in such varieties as 'Flying Cloud' and 'Constellation' is so insignificant that the layman fails to notice it. However, all fuchsia growers would love to see the constitution of such varieties much improved.

The yellow fuchsia is proving much more elusive. There is little among the fuchsia species that would give the breeder much encouragement. *F. procumbens*, the trailing variety, has the most pure form of yellow in it but up to recent times it had proved incompatible with all other fuchsias. A team of scientists from Manchester University did succeed in crossing it with another diploid species *F. splendens*, and promising hybrids did result, but again they proved sterile and so far nothing further has come of this strain.

The shape of the fuchsia has now become so standard in our mind that it is difficult to imagine, say, a fuchsia without a corolla. To many growers it may seem a form of sacrilege to omit from the flower they know what is very often the most beautiful part of the bloom. Nevertheless there is a group of species known as the *Hemsleyella* section of which the type-species is *F. apetala*, and which has only the tube and sepals. Such novelties as might arrive from crossing these with the more conventional species may prove interesting, and may bring forth a feature which many growers might think desirable.

The crossing of another section, the *Encliandra* section, with the larger-type fuchsia has been neglected for far too long. This section, which contains such species as *F. encliandra*, *F. thymifolia* and *F. microphylla*, has the smallest flowers of the whole genus. Some of the species in this section have flowers so small that they have to be really searched for, but they have a delightful shape of their own, with the petals of the corolla opening out flat to the sepals. If the size of the conventional-shaped fuchsia could be combined with the shape of the *Encliandra* section an interesting and fascinating set of hybrids would result. The popular hybrid 'Lottie Hobbie' is one of the few already to come out of this cross, but it is very much '*Encliandra*' still, and needs further development.

There has been, since the end of the First World War, a great deal of speculation as to the worth and possibilities of bi-generic hybrids, not only as it concerns the fuchsia but throughout the vegetable world. So far the scientists have had limited success in this field of experiment, and many an amateur has had as much success with his accidental or chance crosses.

The fuchsia could well benefit if successful bi-generic crossings could be achieved. The family of plants to which the fuchsia belongs is known as *Onagraceae*, which many gardeners prefer to know as the 'evening primrose family', because of the prominent membership of *Oenothera biennis*, the biennial 'evening primrose'. It is not a large family compared with many of the other ones, but it is probably one of the most varied; for beside the *Oenotheras* there are genera such as that well-known aquarium plant *Ludwigia*, herbs such as *Trapa*, the water chestnut, and such well-known annuals as clarkia and godetia. Also among British wild flowers there are two very prominent members of the *Onagraceae* family and therefore near relations to the fuchsia; they are *Epilobium roseum*, the rose willow herb, a perennial which made its name by appearing on all the bombsites of the Second World War and became affectionately known as the 'blitz weed', and *Circaea lutetiana*, the enchanter's nightshade.

Such crosses could well come from *Lopezia coronata*, a hardy annual, the seed of which is easily obtained in packets from certain nurserymen. This delightful annual has spikes of small carmine flowers, the individual flowers having the appearance of tiny butterflies. Furthermore it has 22 chromosomes, as also has *Circaea lutetiana*, the same number of chromosomes as many of the *Fuchsia* species.

The clarkia, godetia and the other members of the family must wait until science has found a way of crossing near relations that have quite different chromosome values.

Bi-generic hybrids are the results of crosses between the genera of one family, in this case *Onagraceae*, and it could well be the answer to any doubtful hardiness of which many fuchsias are suspect, if the extreme degree of hardiness known to be present in the family could be bred into the fuchsia.

Whatever form hybridization may take, the hybridizer must always be on his guard against a breakdown of vigour. It is believed in many exalted circles that a number of the troubles that have befallen such plants as the chrysanthemum, dahlia, and sweet pea have been brought about by the amount of inbreeding and crossbreeding that has been done over the years. Again it has been mentioned that the uses of colchicine and gibberellic acid, and the exposure of the parts of the plants to radiation have done as much harm as good. It may be argued that there is no scientific proof that the treatments mentioned had had any detrimental effects on plant life, but great tragedies have taken place within the human race to prove that drugs cannot be used indiscriminately on any form of life. Their uses are limited and require great care.

Nature has presented the world with a wonderful flower in the form of the fuchsia, and up to this time the hybridizer has added to our pleasure by giving us a wonderful selection to grow. There is still much scope to which the hybridizer can apply his skill and all growers will look forward to whatever worthwhile material he will produce in the years ahead; but all growers will ask that anything that will lower the vigour of the fuchsia, and allow a physical breakdown within the genus, should be destroyed immediately.

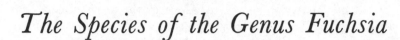

The Species of the Genus Fuchsia

F. abrupta (Johnson 1925) Peru. A rambling shrub with scarlet flowers about 2 inches long. The flowers resemble *F. triphylla* in shape, and are borne in terminal pendent clusters. Has not yet been used for hybridizing, and has nothing further to offer than its type plant, *F. triphylla*.

F. Andrei (Johnson 1935) Ecuador. A small, low-growing species with a green tube suffused red, with the red becoming more prominent in the sepals. The corolla is a purple-red. The flowers are small, only about 1 inch in length.

F. apetala (Ruiz and Pavon 1802) Peru, but spreading into neighbouring states. This is the type species of the section of the genus from which the petals, which form the corolla, are partly or completely lacking. It is described as a vine-like shrub growing up to about 3 feet in height, which gains its support from neighbouring vegetation. The flowers, which are few, are approximately 2 inches long, and consist of an orange-scarlet tube and reddish-orange sepals only. There is no corolla. The flower shape might be a useful addition among the hybrids provided they could be freely produced, and the upright habit of other species could be bred into it. It is not yet available for garden cultivation.

F. arborescens (Suns 1865) Central America. In its native habitat this species makes a small tree, sometimes as high as 25 feet. The small flowers, which have a magenta tube, wine-coloured sepals and a lilac corolla, are borne in erect, well-branched panicles. With its large laurel leaves it is unlike any other fuchsia species and is recognized as a fuchsia only by those who know it.

When in full flower the mass effect of the corymbose panicles of lilac corollas gives the appearance of a lilac spray. For this reason it was known for some time as *F. syringaeflora*. It is recorded that a specimen grown in a cool conservatory in Somerset in 1836 reached the height of 22 feet, with a head of 40 feet in circumference. There is no record that this species has been used successfully in hybridization.

F. asperifolia (Kausc 1905) Peru and Western Brazil. A low-growing sparsely branched shrub with flowers, 1 inch long, growing in dense terminal raceme. The tube and sepals are dark red, and the corolla is scarlet. It is not in general cultivation.

F. Aspiazui (MacBride 1941) Peru. This species is an upright shrub of some 6 feet tall. The flowers, which are about 1½ inches long, are borne in pendent clusters at the end of the branches. The individual flowers have blood-red tube and sepals, and a scarlet corolla. Not in general cultivation.

F. Asplundii (MacBride 1941) Peru. Small shrub which branches fairly freely. Small leaves with red and orange flowers, 2–3 inches in length, which appear singly in the leaf axils. Not in general cultivation but should be studied by the hybridizer.

F. austromontana (Johnson 1939) Peru. An untidy growing, strong, upright shrub with shiny red stems reaching up to 10–12 feet in height. Has a nice medium green foliage which contrasts well with the bright flowers, which are a bright reddy orange colour about 1–1½ inches in length. This is not a very free-flowering species but its colouring might be useful to the plant breeder. Not in general cultivation at the moment.

F. ayavacensis (Kris 1823) Northern Peru and Ecuador. The fuchsia from Ayavaca, a town on the border of Peru and Ecuador. An upright shrub. Not very free flowering for a fuchsia. The colours of the flowers are deep red and scarlet. Not in general cultivation.

F. bacillaris (Lindley 1829) Mexico. A strong growing, densely branched species with small leaves and tiny flowers no longer than a ¼ inch. The tube and sepals are red and the corolla is rose. The sepals reflex back on to the tube on a fully developed flower. Is to be found in catalogues usually under the section headed 'Breviflorae'.

F. boliviana (Roezl 1873) Bolivia. One of the loveliest of the fuchsia species having long large rich crimson clusters of flowers, which show up well against the lightish green leaves. The flowers give way to huge clusters of crimson edible berries. It is a strong-growing shrub, and its only drawback is the susceptibility to damage by frost. Grown as a warm greenhouse shrub it will make a bush of considerable size. Most fuchsia specialists list this species in their catalogues.

F. bracelinae (Munz 1943) Brazil. Nothing was known of this species until Munz published its description in his *Revision of the Genus Fuchsia*. It has very small flowers of the recognized fuchsia combination of red and purple, which are borne in the axils of the leaves. It would appear to have little to commend it either as a garden plant or a subject for the hybridizer.

F. Campos-Portoi (Pilger and Schuize 1935) Brazil. A shrub from the state of Rio de Janeiro, growing in the mountains at heights of 7,000 feet and over. The flowers are small and not particularly plentiful, consisting again of red sepals and a purple corolla. It has an extremely thin leaf most unlike those of other fuchsias. Not in general cultivation and appears to have little value from a horticultural point of view.

F. canescens (Bentham 1845) Colombia. A strong upright shrub growing up to 7 feet or more in its native habitat. The flowers, which are just over 1 inch in length, are described as: 'tube deep scarlet but purplish at base, sepals deep scarlet, corolla bright scarlet'. This species is not obtainable through nurserymen at present.

F. cestroides (Schulze-Menze 1940) Peru. A strong-growing upright shrub which in its native habitat will grow up to 10 feet in height. The flowers, which are about $\frac{3}{4}$ inch long, are dark red in colour and consist of tube and sepals only, the corolla being completely missing. Although this species is not in general cultivation, if there are to be hybrids without a corolla this species may well help the hybridizer.

The Species of the Genus Fuchsia

F. coccinea (Kew 1788) Brazil. This is the species which claims the honour of having been the first fuchsia introduced into England. One of the hardiest fuchsias, it grows into an upright shrub of up to 3½ feet in height. In appearance it is very similar to the more common *F. magellanica* but the flowers are longer. The flower has a red tube and sepals with a purplish-violet corolla.

F. Colensoi (Hooker 1867) New Zealand. This is a freely branching shrub from the North Island of New Zealand. The glabrous leaves are green on the upper surface and whitish underneath. The flowers have a red tube and sepals with a purple corolla. As with all New Zealand species the tube has a peculiar restriction near the base. This species is reasonably hardy in this country but as it is not particularly striking in appearance it is not widely grown.

F. colimae (Jones 1892) Mexico. Botanically placed in the *Encliandra* section of species, the horticultural world still prefer to call it the 'Breviflorae' group. As with all species in this group this one has extremely small flowers. With this species the tiny flowers are almost all white, and are borne singly in the leaf axils. Although many of this group are listed in trade catalogues there is no record of this species being grown in this country.

F. confertifolia (Fielding and Gardner 1844) Peru and West Brazil. Quite a large shrub with extremely small leaves. The flowers, which are very sparse, consist of a dark red tube with a slightly lighter shade for sepals and a lighter red corolla. It is not in general cultivation and is never likely to be.

F. cordifolia (Bentham 1841) Guatemala. This medium-sized shrub which grows in its native habitat to about 3 feet in height has a colour combination within the flower which for a fuchsia is rather unusual. The tube is a dull red and the sepals are green rather more than half-way down from the tips. The corolla is green. The species itself does not branch freely and it does not make an elegant plant. It is, however, obtainable from specialist nurseries.

F. corymbiflora (Ruiz and Pavon 1802) Peru. This is one of the more rampant growing species. The long flowers which have scarlet tubes and sepals are extremely showy, and are borne in large terminal racemes. It does not branch freely and in the days of the large conservatory it was often grown as a climber. The grower who can command heat throughout the year should be proud to have this eyecatching species in his collection. *F. corymbiflora alba* is a variant of this species, having a flower colouring of white tube and sepals with a deep red corolla. *F. corymbiflora alba* is considered by some authorities to be a garden variant introduced by Courcelles (1850), but it has lately been claimed that it is a natural variant discovered by Bentham (1845) growing in Ecuador. Whatever the source of the fuchsia it is well worth growing.

F. Cuatrecasasii (Cuatrecasas 1940) Colombia. A rather poor species with little to commend it as a garden plant. It produces small scarlet flowers very sparsely in terminal racemes.

F. cylindracea (Lindley 1838) Mexico. A member of the *Encliandra* (Breviflorae) section, the tiny flowers are borne solitary in the axils of the leaves, and have a deep red tube and sepals, with a lighter red corolla. It is not in general cultivation but in its native country it is said to grow up to 15 feet high.

F. cyrtandroides (Moore 1940) Mexico. A tall-growing shrub sometimes giving the appearance of being a small tree growing up to 15 feet in height. The small flowers which are not ½ inch long have red tube and sepals with a magenta corolla. It is not in general cultivation at present and appears to have little to offer the hybridizer.

F. decidua (Standley 1929) Mexico. As its name describes this low-spreading shrub is deciduous at the time the flowers are present. Although Munz placed this species in the *Hemsleyella* section, the section of fuchsias which have no corolla, this species does have some extremely small petals but they are so insignificant that it does warrant placing in this class. The flowers, which are about 1 inch long, are borne in short, racemose lateral panicles, and have a vermilion-coloured tube and sepals which rather deepen in colour from the base upwards. It is not in general cultivation.

F. decussata (Ruiz and Pavon 1802) Peru. A shrub up to 10 feet in height in its native country where it is found high in the mountains. The flower is of conventional fuchsia shape and has a red tube and sepals, but the sepals are tipped green. The corolla is a reddish purple. Despite the smallness of the flowers it can be grown into quite an attractive shrub despite its natural lax habit. This species was introduced to this country in 1843 and may well have been used by the early hybridizers although there is little in their introductions to support this. It can be obtained from some specialist nurseries.

F. denticulata (Ruiz and Pavon 1802) Peru and Bolivia. This species is much better known under its synonym *F. serratifolia*. When first discovered by Ruiz and Pavon it was thought that *F. denticulata* and *F. serratifolia* were two different species, but it has now been resolved that the difference is merely colour variation within the species *F. denticulata*. This is, from a garden point of view, one of the loveliest of the species. Grown in a greenhouse border it will make a large plant. The leaves are among the largest in the genus and have a distinct, metallic green above and a much paler green below. The flowers have a long tube which is a deep pink at the base becoming paler towards the sepals which have a most distinct green tip. The corolla is red. It is slightly tender but will flourish in the cool greenhouse. It is catalogued by most fuchsia specialists under its synonym. It has been used in hybridizing, and that lovely hybrid 'Fanfare' is one of its better-known progeny.

F. encliandra (Stendel 1840) Mexico. The type species of the section of the genus which bears its name. Obtainable from certain nurseries but may be listed under its synonym *F. parviflora*. Like most of the 'Breviflorae' fuchsias this species is a rampant-growing shrub with the tiniest of flowers which are red. The leaves are very small and a well-grown shrub, from a distance, has the appearance of being a fern.

F. excorticata (Forster 1776) New Zealand. Not a striking species but certainly an interesting one. In New Zealand it grows to tree-like proportions with a wood that has an attractive grain, and which is used by the Maoris to make small articles for the tourist trade. In the counties of Devon and Cornwall quite large specimens have flourished in the open for many years, but its hardiness is suspect in other parts of the country. The leaves of this species are of a lightish glossy green on the upper surface and have a glaucous, almost white appearance below. The flowers when fully developed are distinct in shape, the purple-red tube having an unusual restriction near its base. The corolla is of a very deep purple which from a distance looks almost black. Probably the first thing about the flower which catches the eye is the beautiful blue pollen on the anthers. This is a most striking feature and shows up well the large round yellow stigma.

F. Fischerii (Macbride 1941) Brazil. One of the last species to have been discovered. Very little is known about it except that it has a rather lax, untidy habit, and the inch-long flowers have a pinkish tube and sepals with a purple corolla.

F. fulgens (de Candolle) Mexico. This is indeed a true aristocrat of the genus. Its light sage-green, hairy leaves are the largest of any known fuchsia, often some 9 inches long and 6 inches wide. Its flowers are pendulous and grow at the extremity of the branches in large, long clusters. The tube is long and cylindrical, starting very thin at the ovary and widening all the way down to the sepals, and is of a light vermilion-red. The triangular sepals are short, starting with the tube colour at their widest part, and finishing with a distinctive green tip. The corolla is vermilion, but brighter than the tube, with oval petals pointed at the tips, and these are a little shorter than the sepals. This species has proved accommodating in every way since its introduction. Up to the time of the introduction of *F. fulgens* the species that were being used for hybridization were small and of very limited colour range. *F. fulgens* with its much larger flowers and its unusual colourings was just what was needed to create renewed interest in the genus, and the garden varieties that came out of it were distinctive and immediately popular. The green-tipped sepals seen on many of the garden varieties listed today are a legacy of those early crosses made with this species as one of the parents. Planted in the greenhouse border *F. fulgens* will make a bush of handsome proportions and a specimen to be seen in the Large Temperate House at Kew Gardens is a fine example of this form of cultivation. As a pot plant it is ideal for exhibition purposes and sure to catch the judges' eye. There are four known natural variants of the species and they are as follows:

F. fulgens carminata rosea. Identical in habit as the type but the tube is a rosy carmine in colour.

F. fulgens gesneriana. Not so upright as the type, and the tube is not quite so long and more squarish in appearance.

F. fulgens multiflora pumila. Dwarf and very floriferous in habit although the flower lacks the length of the type species.

F. fulgens rubra grandiflora. Very similar in habit to the type species but the flowers have a very long tapered tube. Perhaps the most elegant-looking flower in the genus.

F. furfuracea (Johnson 1925) Bolivia. A low-growing, spreading shrub which bears its few flowers in terminal racemes. The 1½-inch long flowers are almost self-coloured red, although a faint trace of purple can be seen in the corolla. This species is not in general cultivation, and appears to have little to commend it.

F. Garleppiana (Kuntze and Wittmack 1893) Bolivia and Peru. Epiphytic in habit, this species belongs to the group of species in which the corolla is completely lacking. It has tuberous roots and the pink flowers are sparse. It is not in general cultivation.

F. Gehrigeri (Gehriger 1930) Venezuela. A scrambling shrub which will climb over rocks and adjacent vegetation sometimes to a height of 15 feet. It bears its few flowers in pendulous, terminal clusters. In colour the 2-inch flowers have a dark red tube with lighter sepals, and dark red corolla. It is not in general cultivation.

F. glaberrima (Johnson 1925) Ecuador. The flowers have a bright red tube with scarlet sepals and a scarlet corolla. These are borne in short terminal racemes on a sturdy upright growing shrub. Not in general cultivation.

F. Hartwegii (Bentham 1845) Colombia. A large upright shrub bearing numerous ¾-inch-long flowers which are borne in pendulous terminal clusters. The flowers are described as having red tube with orange-red sepals, and red corolla. Although this species is not in general cultivation it would appear to be quite an acquisition to the collector.

F. Hemsleyana (Woodson & Seibert 1937) Costa Rica and Panama. A tall-growing shrub of the 'Breviflorae' section. The minute flowers are completely rose in colour. Like all the species in this section, it is dainty and delightful in growth. This species is obtainable but may take some looking for.

F. hirsuta (Hemsley 1876) Peru and Brazil. A tuberous-rooted epiphyte, bearing bright scarlet flowers having long funnel-shaped tubes and no corolla. It can be found in its native habitat growing high in trees, or being suspended on the rock-faces of the mountains. It has an almost distinctive leaf as its name implies, being completely covered both on the upper and lower surfaces by fine hairs. Not in general cultivation.

F. hirtella (Kris 1823) Colombia. A low-growing shrub. The flowers, which are borne in terminal, pendent clusters, are approximately 1½ inches long, have a rose-red tube and sepals, with a reddish-scarlet corolla. It is not in general cultivation.

F. hypoleuca (Johnson 1925) Ecuador. An upright-growing shrub bearing flowers about 1 inch long, which have red tubes with scarlet sepals and corollas. Not in general cultivation.

F. Jahni (Munz 1943) Venezuela. A new species first described by Munz in his revision of the genus. The red flowers would appear to be very small and extremely sparse. The shrub is reputed to be of scrambling habit. Seems to have very little to commend it, and is not in general cultivation.

F. juntasensis (O. Kuntze 1898) Bolivia. Known as the 'vine' fuchsia, it is another epiphyte. In its native country it grows in thickets and on rock-faces. The 2-inch-long flowers have a pink tube with sepals which are of a deeper shade. There is no corolla. The sepals open well and at times are slightly recurved. The flowers are borne in terminal clusters and their numbers vary considerably according to the locality. Although not in general cultivation this could prove an interesting species to the hybridizer. The main difficulty would appear to be copying the conditions of its native habitat in order to get the best results from the shrub.

F. Killipii (Johnson 1928) Colombia and Venezuela. A tall upright shrub which gains much support from surrounding vegetation. The large number of 1½-inch flowers are borne in terminal racemes, and have red tubes and sepals, with a brighter corolla. Although this would seem to be worthy of collection it is not yet in general cultivation.

F. Kirkii (Hooker 1868) New Zealand. A woody, creeping species whose flowers point skywards in a most unfuchsia-like way. The small flowers have a tube which is dull red shading to greenish yellow, with sepals that are green tipped purple. This species is almost identical with its more famous neighbour *F. procumbens*, the difference being that *F. Kirkii* has a wider tube. This species is available, although it will have to be hunted and the grower will have to be sure that his plant does not turn out to be *F. procumbens*.

F. Lehmannii (Munz 1943) Ecuador. Another separate species created by Munz in 1943. It is a many-flowered species, the tube is red, and the sepals and corolla scarlet. The flowers are borne in terminal racemes. It is not yet in general cultivation.

F. leptopoda (Krause 1905) Peru. A strong upright shrub bearing flowers which have a dark red tube, rich red sepals, and glowing red corolla. The flowers appear singly in the upper leaf axils, and they are well shown up by the lightish green leaves. This species may well be in cultivation in this country, although it is not catalogued. It is definitely grown in California where it has occasionally been used for hybridizing. The variety 'Fanfare' is a result of the cross *F. denticulata* × *F. leptopoda*, and the rich colouring for which it is noted is definitely from *F. leptopoda*.

F. Llewelynii (Macbride 1941) Peru. A bushy species growing up to 3 feet in height. The flowers, which are sometimes almost 2 inches in length, are sparse, and have a red tube and sepals with a violet corolla. It is not in general cultivation.

F. loxensis (Kris 1823) Peru. Named after the neighbourhood of Loxa from whence it hails, this species was recorded as having been in cultivation at Kew in 1862. From there it was described as a shrub of dense, leafy habit, with flowers about 1 inch long, and deep red in colour. It now appears to have gone from general cultivation.

F. lycioides Chile. This is one of the earliest species ever collected, having been recorded as being in the king's garden at Kew in 1796. It is quite hardy in this country but grown in the open it has little to commend it. Only in a very warm summer will it grow well and produce any profusion of flowers. Pot-grown with a very restricted root run, and where conditions can be controlled, it does produce a vast number of its flowers, which are delightful, with the red tube and sepals and the purplish red corolla. Grown under such conditions it can be a mass of colour.

F. macrantha (Hooker 1845) Peru. This beautiful species belongs to that section of the genus which has no corolla. It has large flowers and when it was shown at the Horticultural Society's rooms in 1846, by the famous nurserymen James Veitch & Sons, they were described as being 6 inches long. It is epiphyte in habit and in Peru grows on the mountain trees. The flowers are borne in clusters at the end of the shoots, and have a tube of soft rosy red, passing into yellow and green on the edges and tips of the sepals. It has unfortunately passed out of cultivation temporarily.

F. macrophylla (Johnson 1925) Peru. A strong upright bush with stems of a red or purple tinge. The flowers of about $\frac{3}{4}$ inch long have a scarlet tube, red sepals with green tips, and the corolla is bright red. It is not in general cultivation.

F. macrostigma (Bentham 1844) Ecuador. This large-flowered species is known to grow strongly up to 5 feet in its native habitat. The $\frac{3}{4}$-inch-long flowers are solitary in the upper leaf axils, and have a purple-red tube, with paler sepals, and a corolla cerise to crimson. The subspecies *F. macrostigma longiflora* is probably more choice than the type species. The latter would appear to be an acquisition to any collector of fuchsias.

F. magdalenae (Munz 1943) Colombia. In appearance this shrub very much resembles the species *F. denticulata* except that the flower has a tube which is purple at the base, changing to red as it reaches the sepals. The sepals are red and the corolla scarlet. It is a species that specialists would welcome in cultivation.

F. magellanica (Lamarck 1768). This is the type species of a great family. The flowers consist of a red tube with a purple corolla. It is completely hardy in this country, and in all its forms it can be planted out with confidence. This species comes from the southernmost part of South America where it flourishes in weather conditions often worse than those experienced in the British Isles. Both the type and the subspecies have a vast number of uses in the garden. In some parts of these islands it has naturalized itself, and together with the F1 hybrid, F. Riccartonii, can be seen forming glorious hedges of rich red and purple.

This species can proudly claim that the majority of the present-day hybrids have *F. magellanica* in their family tree. Extensively used by the early hybridizers it has imparted a great deal of its hardiness into what are now known as the semi-hardy varieties. Even in Nature *F. magellanica* has hybridized with other species and the following are the results:

F. m. var. *molinae* (synonym *m. alba*) tube white, sepals and corolla very pale lilac.

F. m. var. *americana elegans*. Smaller flowers with brighter coloured tube.

F. m. var. *conica*. More rounded buds.

F. m. var. *discolor*. Dwarf habit.

F. m. var. *globosa*. Much heavier flower than the type.

F. m. var. *gracilis*. Very slender arching growth and a very rampant grower.

F. m. var. *gracilis variegata*. As above but with a lovely silvery variegated foliage.

F. m. var. *longipedunculata*. Long slender blooms.

F. m. var. *pumila*. A tiny plant with tiny flowers. Ideal for rockery.

F. m. var. *Thompsonii*. Brighter colours than the type.

All the varieties are worth growing in the garden but *F. m. molinae* will only produce a real profusion of flower if it is given a very restricted root run. In these conditions it is a wonderful sight in full flower.

F. membranacea (Hemsley 1876) Ecuador and Venezuela. A little-known species. The flowers, about $1\frac{1}{2}$ inches long, have a greenish tube suffused with red. The corolla is completely lacking.

F. Mexiae (Munz 1943) Mexico. A little-known shrub first described by Munz in his revision of the genus. A member of the 'Breviflorae' section it has a tiny flower which has a red tube and sepals and a white corolla. Like the rest of its section it grows into quite a large bush.

F. michoacanensis (Sessé and Mocino 1887) Costa Rica and Mexico. Another tiny-flowered species of the 'Breviflorae' family and having the characteristic strong upright growth of that group. The flowers are borne singly in the axils of the leaves and have red tube and sepals with a coral corolla. This species can now be found in some specialist catalogues.

F. microphylla (Kris 1823) Mexico. A densely branched shrub of the 'Breviflorae' section. An upright habit, but not the vigour of most of this section, it has a delicate fern-like appearance when seen from a distance. The tiny flower has a deep red tube and sepals, with a rose corolla. It is obtainable from specialist nurserymen.

F. minimiflora (Hemsley 1880) Mexico. A member of the 'Breviflorae' section. A strong upright-growing shrub having the tiniest flower of the whole genus. The flowers, which are sometimes less than ⅛ inch long, are whitish and suffused with red. It is charming to growers who like their charm in miniature.

F. minutiflora (Hemsley 1878) Mexico and West Indies. Another of the 'Breviflorae' section. A smaller-growing shrub than most of its near relations. The growth is thin and wiry in appearance giving the plant a fernlike appearance. The flowers are very small, less than ¼ inch in length. The tube and sepals are red and the corolla white. This species is obtainable from some specialist nurseries.

F. Munzii (Macbride 1941) Peru. A species of which very little seems to be known. It would appear to bear but few flowers which are in terminal panicles, the tube and sepals being red, and the corolla is also red with a trace of purple in it. It does not seem to have anything in its make-up which would appeal to the gardener.

F. Osgoodii (Macbride 1941) Peru. A strong, upright shrub which in its native habitat will grow strongly up to 12 feet in height. The flowers which are borne in a few racemes are just over 1 inch long. The tube is a dark red, and the sepals and corolla are red, but lighter than the tube. It is not in general cultivation.

F. ovalis (Ruiz and Pavon 1802) Peru. A bushy species growing up to 3 feet in height. It is almost a self-coloured flower, the tube and sepals being scarlet, and the corolla also scarlet but shading to purple. Flowers are borne in small racemes coming from the axils of the upper leaves. The leaf is of quite a dark green. There is no record of this species being in general cultivation.

F. pallescens (Diels 1938) Ecuador. A medium-sized shrub about which very little is known. The inch-long flowers have a pale carmine tube, white sepals and purple corolla. This species may prove more interesting when more knowledge is forthcoming.

F. perscandens (Cockayne and Allen 1927) New Zealand. The New Zealand climbing species. The flower is similar to *F. Colensoi*, having the same restriction in the tube and having the same colourings—tube red, sepals greenish red, and corolla purple. The seed fruit is a large dark purple berry. It is not in general cultivation.

F. petiolaris (Kris 1823) Colombia. A shrub 6 feet tall bearing its flowers in the upper leaf axils. The flowers have a red tube and sepals with a darker red corolla. They are about 3 inches long. This species is not in general cultivation although it must have been at one time as it is illustrated in *Flore des Serres* t. 481 under the synonym *F. miniata*.

F. pilosa (Fielding and Gardner 1844) Peru. A low-growing shrub with few branches. The foliage is slightly hairy, and the flowers are borne in terminal racemes. The flower is almost a scarlet self-colour about 1 inch long. Not in general cultivation.

F. platypetala (Johnson 1939) Peru. Tall, erect-growing shrub. The flower is distinctive, having a red tube, crimson sepals and crimson petals, each with a white blotch in the middle, forming the corolla. These are borne in the upper leaf axils. The leaves are covered in very fine hairs and the young wood is very reddish in appearance. Such a species may well find some use in hybridizing when eventually brought into general cultivation.

F. polyantha (Killip 1935) Colombia. A small shrub which has flowers with a purplish-red tube, scarlet sepals and a crimson corolla. These are plentiful and are carried in terminal, pendulous panicles. The leaves are light green in colour. An interesting species that fuchsia specialists will welcome into general cultivation.

F. Pringlei (Robinson and Seaton 1893) Mexico. Another 'Breviflorae' fuchsia. A small shrub bearing tiny flowers which are pale pink in colour. This may be in cultivation although it is not catalogued in this country.

F. Pringsheimii (Urban 1898) San Domingo–West Indies. A very upright shrub. The flowers are borne in the axils of the leaves and are almost a red self-colour, although there is a faint shade of green in the sepals. Not in general cultivation.

F. procumbens (Cunningham 1839) New Zealand. An unusual, hardy species, which scrambles over the ground rooting as it spreads in much the same way as the common periwinkle, although nothing like as rampant or invasive as that plant. The flowers, in a most unfuchsia-like way point upwards, are small with an orange-yellow tube, green on the outside of the sepals with a reddish-purple colouring on their upper surface. The seed pods, or fruit, are quite large and look like small red plums. An interesting plant to grow and worthy of a place in any fuchsia collection.

F. putumayensis (Munz 1943) Colombia. A little-known species. The flowers are borne in compact terminal racemes, and have a bright red tube and sepals with a scarlet corolla. Not in general cultivation.

F. regia (Gardner 1842) Brazil. Usually catalogued under its synonym *F. alpestris*. A rampant species climbing to a height of 20 feet or more, with soft pubescent leaves and very small slender flowers with red tube and sepals, and corollas of slightly darker shade. The branches are long and slender. Crossed with the hybrid 'Royal Purple' it brought forth the climbing variety

'Lady Boothby' which has put up such a grand display in the Temperate House at Kew. The type has a natural variant *F. regia* var. *affina* which has less hairy leaves. For those who have a large greenhouse this species makes an ideal climber.

F. rivularis (Macbride 1940) Peru. Very little is known about this species except that it has solitary flowers growing from the upper leaf axils. The flower colourings are: tube and sepals dark red, with a corolla of purplish red. The habit of growth and the general appearance of the shrub are said to be similar to those of *F. canescens*. Not in general cultivation.

F. salicifolia (Hemsley 1876) Bolivia. This is another of the epiphytic fuchsias. Like the others with this habit, the flower has no corolla. The flower, which is about 2 inches long, is yellowish green in colour. Its growth can be described as shrubby. Although this species is not in general cultivation the dried specimens sent to Kew by botanists from Bolivia give every indication that this is a handsome species.

F. Sanctea-Rosea (Kuntze 1898) Bolivia and Peru. A fairly strong-growing shrub. It bears numerous flowers which have a bright red tube with scarlet sepals, and a bright orange-red corolla. The flowers are borne in the upper leaf axils, and as it is a very leafy shrub it bears many flowers. Although not an easy species to grow, the sight of it in full bloom is very rewarding. Only specialist nurserymen list this shrub.

F. scabruiscula (Bentham 1845) Ecuador. A species which seems to have very little to commend it. The inch-long flowers are borne singly in the upper leaf axils on a low spreading bush. It is not in general cultivation.

F. sessilifolia (Jameson 1835) Colombia. A large shrub or small tree in its native country. The flowers are borne in clusters at the ends of pendant shoots, they are 1 inch long with scarlet tube and spreading yellowish sepals. The corolla is scarlet. Although not now in cultivation it is recorded that this species was raised from seed in Edinburgh in the year 1865. It was shown in full flower by a Mr. Anderson Henry. It is to be hoped that someone will reintroduce this species.

F. simplicicaulis (Ruiz and Pavon 1802) Peru. A beautiful species which first flowered in this country in 1858. The tube, sepals and corolla are a bright purplish red, the flowers being from 2½–3 inches in length and produced in clusters at the ends of the drooping shoots. For many years this species was a feature of the No. 4 Temperate House at Kew, until it was pulled down in 1962. It grew up to the roof and was trained under the roof glass. It is to be hoped that in the new house *F. simplicicaulis* will again take its place.

F. Skutchiana (Munz 1943) Guatemala. A member of the 'Breviflorae' section, a strong upright shrub bearing extremely minute flowers. The flowers are a whitish pink in colour, and they turn red as they age. It is in cultivation in this country.

F. Smithii (Munz 1943) Colombia. A loosely growing shrub which relies on surrounding vegetation for its support. The flowers are approximately 2½ inches in length with a pinkish-red tube, dark red sepals, and a very dark red corolla. It is not in general cultivation.

F. splendens (Zuccarini 1832) Mexico. A very distinct species. First found its way into cultivation through the seed collected by Mr. Hartweg in 1845 on Mount Totonoepeque, at an altitude of 10,000 feet above sea-level. It is a loosely branched shrub growing up to 8 feet in height. The short tube is rose to bright red, while the short and almost erect sepals are green, with a tinge of red in them. The corolla is pale green. Kew has several fine bushes in the large Temperate House, although planted as they are in the borders they do not produce the wealth of flower that pot-bound specimens will give. A fine species.

F. Storkii (Munz 1943) Peru. A strong upright shrub. The 1½-inch-long flowers are borne in terminal panicles and are dark red in colour. It is not a widely known species and is not in cultivation.

F. striolata (Lundell 1940) Mexico. A member of the 'Breviflorae' section which is more lax in growth than most of this section. The tiny flowers—they are only ¼ inch long—are borne solitarily in the leaf axils, and are red in colour. It appears to have nothing to offer the gardeners or hybridizers.

F. sylvatica (Bentham 1845) Colombia. A low-growing, almost trailing shrub which bears its flowers in terminal pendent racemes. The inch-long flowers have a red tube and sepals, while the corolla varies between crimson and purple. It has nothing which the hybridizer has not already used.

F. tacanensis (Lundell 1940) Mexico. Another 'Breviflorae' species with tiny flowers growing on a strong upright bush. The flowers arc solitary in the leaf axils and are almost white when first out, and as they age they turn pink to red. Many of this section are in cultivation and this one may well be among them.

F. tetradactyla (Lindley 1846) Guatemala. Of the same section as the previously described species, although not so vigorous in growth. The very small flowers have a red tube and sepals and a rose-coloured corolla.

F. thymifolia (Kris 1823) Mexico. Also of the 'Breviflorae' section. A shrub of medium growth, it bears tiny flowers which arc whitish, turning pink to red with age. There has been a great deal of confusion in recognizing the species in this section as the colours vary slightly according to the conditions and environment, and there is normally so little to choose between many of them that the slightest variation becomes misleading.

F. tincta (Johnson 1939) Peru. A low shrub of about 4 feet in height. The flowers are borne in terminal racemes, and are about 1½ inches long. They have a deep crimson tube with slightly lighter sepals, and a crimson-scarlet corolla. It is not in general cultivation.

F. Townsendii (Johnson 1925) Ecuador. A rather sprawling low-growing shrub which has a very distinctive dark purple wood. It has flowers which have a red tube suffused with green, red sepals, and a scarlet corolla. These are borne singly in the leaf axils and are not very profuse. It is a species of which very few details are known.

F. triphylla (Plumier 1703) San Domingo. This is the species on which the genus *Fuchsia* was founded. It is quite distinct among a very varied genus with its brilliant bright red to scarlet flowers, growing on a shrubby erect plant. The leaves have a copper-bronze appearance which enhances the vivid colour of the flowers. It is considered a difficult plant to grow under the artificial conditions it meets in this country, and it is the one fuchsia that is extremely frost-shy. It has been quite extensively used in hybridization and its progeny popularly known as the 'Triphylla hybrids' are among the most beautiful of the fuchsia varieties.

F. tuberosa (Krause 1905) Peru. An epiphyte with an unusual fuchsia flower. The red tube is swollen in appearance and the sepals are entirely green. It has no corolla. These flowers are borne in the upper leaf axils. This species has been cultivated by growers on the Pacific coast of the United States, where it has the reputation of being extremely temperamental. So far all attempts to cross this shrub with the more conventional types of fuchsia have been unsuccessful. To date *F. tuberosa* has not been introduced into the United Kingdom.

F. tunariensis (Kuntze 1898) Bolivia. Like the previously described species, *F. tunariensis* has no corolla on the flower. Also epiphytic in habit, it has thick tuberous roots. The few flowers which it bears are carried in the upper leaf axils. They are 2 inches long and have red tubes and sepals. It is not in general cultivation.

F. unduavensis (Munz 1943) Bolivia. Another epiphytic, tuberous-rooted species, which in its native habitat scrambles over rocks and adjacent vegetation. It has 2-inch-long flowers with bright purplish-red tubes and red sepals. There are no petals to make up the corolla. Not in general cultivation.

F. venusta (Kris 1823) Colombia. This species was grown quite extensively in the large Victorian glasshouses, but now appears to have vanished from cultivation. It is an upright, rather fragile-looking shrub bearing its flowers in terminal pendent corymbose raceme. The individual flowers have a red tube and sepals, with a carmine corolla, and are about 3 inches in length. It is a handsome species and fuchsia enthusiasts would welcome its reintroduction.

F. verrucosa (Hartweg 1845) Colombia. A strong, upright-growing shrub which is rather shy to flower. The few flowers that it bears are quite small and are borne singly in the upper leaf axils. They are almost red self in colour. It appears to be a species which has little to offer as a garden plant.

F. Woytkowskii (Macbride 1941) Peru. A strong, upright-growing shrub. The flowers are solitary in the upper axils and have a deep vermilion tube with lighter sepals, and a bright red corolla. They are approximately 2 inches long.

Note to the Chapters which Follow

The exact number of varieties of hybrid fuchsias that have been raised and marketed is not known. Furthermore the number of varieties still in cultivation is an unknown factor, although it is known that there are several thousand different varieties scattered throughout the world.

The International Society for Horticultural Science, centred on the Hague, did, in the year 1967, take the first steps to record the introduction of the new varieties by appointing the American Fuchsia Society the International Registration Authority for Fuchsias. By this appointment the American Fuchsia Society becomes the official registrar of all new fuchsia introductions throughout the world. All gardeners who have any interest in the flower will welcome this appointment as perhaps the first step toward sorting out the confusion that exists in the nomenclature of fuchsia cultivars. At the present it can only hope to prevent the confusion from worsening, for until some authority has the time, the money, and the know-how to undertake registration, and, more important still, the identification of the existing varieties, there must inevitably be bewilderment and disagreement among the enthusiasts and gardeners generally.

The following chapters are not intended to be a fully comprehensive 'check list' of fuchsias, but a selection of those that are readily available from specialist nurserymen. They are divided into separate sections to help the reader to understand what certain varieties are capable of, rather than being a rigid guide as to the manner in which they should be grown.

Every gardener is an individualist, and as such, can only have ideas suggested, rather than dictated to him. These lists then are only suggestions.

CHAPTER XII

Varieties Suitable for House
and Window Sill

'Athela' (Whiteman 1942). Single. Tube and sepals creamy pink, corolla salmon-pink, deeper shade at base. Raised by the first Secretary of the British Fuchsia Society, this is a most accommodating variety. Holds its flower well in the house, and is capable of withstanding the stringent conditions of window-box growing. Suitable also for summer bedding, and in the greenhouse will make a fine standard.

'Bengali' (Rozain-Boucharlat 1911). Double. Tube and sepals scarlet, corolla purple but suffused and edged scarlet. The medium-sized flowers are profuse and of good shape. The sepals are reflexed, revealing the beauty of the petals. The plant has a low spreading habit very suitable for the window sill.

'Berliner Kind' (Eggebrecht 1882). Double. Flowers of medium size and very double. Scarlet sepals with pure white corolla. The flowers are very freely produced on a low-growing bush.

'Bouquet' (Rozain-Boucharlat 1893). Double. Sepals coral-red, corolla violet-purple. A quite hardy variety which produces its small flowers very profusely. Of dwarf habit it is ideal for the window box. Not very popular.

'Display' (Smith 1881). Single. Tube and sepals rose-pink, corolla deeper pink. One of the best all-round fuchsias, and in the greenhouse is capable of being trained to all the recognized shapes. However, it does hold its flower remarkably well under the conditions met in the average house, and for this reason is included in this section. The flowers are of distinctive shape, having a saucer-shaped corolla, and they are borne in great numbers. Very popular and worthy of that popularity. Also good for summer bedding, and in the greenhouse can be grown as a standard.

'Dollar Princess' (Lemoine 1912). Double. Better known in some areas as 'Princess Dollar'. Tube and sepals cerise, corolla rich purple. A vigorous and extremely free-flowering variety. This variety is suitable also for growing in standard and pyramid form. It produces bloom early in the season and owes its popularity to the ease with which it can be grown. Most florists and flower-market stalls offer this one for sale some time during the season. Although the flowers are small for a double it makes up for the lack of size by the number of flowers produced.

'Flash' (Hazard). Single. Tube, corolla and sepals light magenta. The flowers are small but are borne in profusion. The low spreading habit of the variety makes it ideal for window-box growing.

'Grus aus Bodethal' (Gebruber-Teupel 1904). Single. The tube and sepals are a bright crimson and the corolla opens so dark a purple at first that it looks almost black. The corolla turns a rich purple with age. The flowers are of medium size and freely produced. An eye-catching variety.

'Lady Thumb' (Roe 1966). Single. A sport from the well-known hardy variety 'Tom Thumb'. Habit and shape of blooms identical to that of its parent. New colouring is: tube and sepals bright red. Corolla white. A useful addition to the hardy section.

'Liebreitz' (Kohene 1874). Single. Tube and sepals are pale cerise and the petals of the corolla are so veined as to give them a pale pink appearance. The flowers are quite small but are borne in great profusion on a dwarf, bushy, and most compact plant. This variety has long been lost to cultivation, but its rediscovery and reintroduction has been welcomed by enthusiasts, and it is immensely popular. It is quite hardy, and when grown for exhibition can be a veritable ball of flower.

'Miniature' (Lemoine 1894). Single. Tube and sepals cerise, corolla purple. Not an easy variety to obtain but is quite hardy and will survive in the most exposed window box. The flowers are small and very free. Properly grown will make a tough little bush.

'Miss Prim' (Reiter 1947) Semi-double. Tube and sepals carmine, corolla deep imperial purple, changing to lighter purple towards base of petals. A very free-branching variety which produces its flowers on every branch. A very easy plant to grow and makes a delightful bushy plant.

'Mr. A. Huggett' (Raiser and date unknown). Single. Tube and sepals cerise, corolla mauvish pink. The small flowers are produced very freely and the upright branches need little tying. An extremely attractive variety which grows well under all conditions. One that will put up with a certain amount of neglect.

'Mr. W. Rundle' (Rundle 1896). Single. Tube and sepals pale rose, corolla orange-vermilion. A variety that is often seen in the local markets. A medium-sized flower produced on upright branches. Holds on to its flowers well indoors and is capable of standing window-box conditions. An excellent variety. Will make a fine standard in the greenhouse, or bedded out.

'Mrs. Ida Noack' (Lemoine 1911). Single. The waxy rose sepals overhang the magenta corolla. The flowers are small but profuse. Of bushy habit it will stand a lot of maltreatment. Very easy to grow.

'Mrs. Victor Reiter' (Reiter 1940). Single. The tube is creamy white in colour and very long. The sepals are of the same colour, thin and pointed. The single corolla is of rich crimson, except at base, where the colour alters to creamy white. The largish flowers are borne on lax branches which arch gracefully, sometimes cascading. An outstanding variety to grace the window box. Apart from the window box it is eminently suitable for basket, bush, or half standard.

'Nina Wills' (Wills 1961). Single. This sport of the popular variety 'Forget-me-not' has pale flesh-coloured sepals with soft baby-pink corolla. In every other way it is identical to the more famous parent, and like its parent is well worth growing.

'Other Fellow' (Hazard 1946). Single. Tube and sepals waxy white, corolla coral-pink, shading to white at base. The many flowers are produced on a bushy upright shrub. The whole appearance of the variety is one of daintiness. An excellent variety.

'Peter Pan' (Erickson 1960). Single. The tube and upturned sepals are a very deep pink. The corolla is orchid-purple with splashes of lilac here and there. The flowers are small but produced in quantity on a low spreading bush. Has proved that it can stand the rigours of window-box culture and has so far proved quite hardy in the south of England.

'Petite' (Waltz 1953). Double. A very heavy-blooming variety producing its small flowers on an upright bushy shrub. The tube and sepals are pale rose, the corolla lilac-blue, fading to lavender-blue. It flowers early in the season and maintains a profusion of bloom throughout the summer. An easy variety which is ideal for the window box.

'Pink Bon Accorde' (Thorne 1959). Single. Tube and sepals are a pale pink and the corolla is a really deep pink. Flowers are smallish but grow in vast numbers. Growth is upright, short jointed, and needs no support. A really first-class variety which is said to be a sport of the upright standing variety 'Bon Accorde', but it is difficult to see even the slightest resemblance. Holds its flowers well indoors.

'Pink Delight' (Brown Soules 1957). Double. Tube and sepals rose-pink, corolla delicate soft pink. The medium-sized flowers are quite free. This variety requires very little training as it branches freely and evenly of its own accord. The result is a compact little bush which everyone will admire.

'Pink Jade' (Pybon 1958). Single. Tube and sepals pink shading to green at the tips, corolla orchid-pink with a picotee edge of rose. The corolla is saucer-shaped in much the same manner as 'Display'. The shrub is short-jointed and rather slow-growing but it produces its attractive flowers quite freely. Very aptly named.

'Powder Blue' (Niederholzer 1947). Semi-double. Tube and sepals rose, corolla pale blue. The flowers are of medium size, freely produced. Grown in a shaded greenhouse it will grow quite tall and upright, but in a window box which does not get a great deal of sun it will become more bushy in appearance and will flower where many others would fail.

'President Roosevelt' (Garson 1942). Double. Tube and sepals coral-red, corolla dark purple. The flowers are smaller than most doubles but grow in great numbers. It grows as a shrubby bush and besides being ideal for the window box it will hold its flowers quite well in the house itself.

'R.A.F.' (Garson 1942). Double. Red tube and sepals, corolla powder pink. This variety is constantly being compared with that very popular variety 'Fascination', but it is more bushy in habit and the flowers, while being of the same colour and shape, are smaller, and slightly more freely produced. It is a handsome plant when in full bloom and one which is always admired. Away from the window box or house it will make a fine half standard.

'Scabieuse' (Rozain-Boucharlat 1928). Double. A medium-sized flower with red tube and sepals. The corolla is made up of outer petals which are white, suffused blue, and the inner petals which are purple. An upright-growing shrub which holds its flowers well in the house.

'Scarcity' (Lye 1869). Single. The tube and sepals are cerise and the corolla is a rich rosy purple. Flowers are of medium size and are produced in great profusion. The flowers are quite distinct, being of rather stout appearance, and they appear even under the most adverse conditions. An excellent variety which has proved itself quite hardy in many parts of the country. Will make a grand standard in the greenhouse.

'Silver Queen' (Haag 1953). Semi-double. A very early free-flowering variety, with flowers having deep rose tube and sepals, with a silver-blue corolla. The flowers are not large but they are plentiful, and they grow on a neat compact bushy shrub. A good variety.

'Strawberry Queen' (Haag 1937). Single. Red tube and sepals with a deep strawberry-red corolla. The flowers are small but profuse. It is almost completely hardy in this country and grown in a window box it will take all the rough treatment given to it. Under such conditions it will make a neat little shrub.

'Temptation' (Peterson 1959). Single. The tube and long sepals are white. The petals of the corolla are orange-rose with white throats. The flowers are of medium size and are borne in great profusion. The leaves are long and distinctly crinkly. Growth is strong and upright. A variety suitable for all types of growth. A really worthy fuchsia.

'Tom's Chum' (Catt 1951). Single. Said to be a seedling from the more famous variety 'Tom Thumb', to which it is very similar in habit although different in colouring. The tube and sepals are crimson and the corolla purple. The colours are generally darker than those of its parent. Hardy and dwarf in habit, it grows well under window-box conditions.

'Wave of Life' (Henderson 1869). Semi-double. Tube and sepals scarlet, with purple corolla. The foliage is an attractive golden-green. The flowers are small but quite freely produced. It is of a low spreading habit and is useful both for house and window box.

CHAPTER XIII

Varieties Suitable for Hedges

'**Caledonia**' (Lemoine 1899). Single. Long reddish-cerise tube and sepals, with reddish-violet corolla. Extremely hardy variety of bushy habit. Its branches are upright, slender and very graceful. Free flowering with palish foliage of an unusual rough-looking texture. It is quite a distinct variety. Rather dwarf in habit, it makes a delightful hedge along the top of a retaining wall where the growth will completely disguise any harshness.

'**Corallina**' (Pince 1843). Single. Tube and sepals bright scarlet, corolla deep rich purple. For a hardy variety the flowers are certainly large and very freely produced. The greenish-bronze foliage adds further to the beauty of the variety. Strong-growing but rather lax in habit. Will make an attractive hedge if given a supporting wire a short distance from the ground, so that the lower branches are allowed to arch over.

'**Dorothy**' (Wood 1946). Single. Tube and sepals are bright crimson, and the corolla is violet, veined red. The flowers are of a hardy variety of medium size and freely produced. It has a vigorous upright growth and is capable in a sheltered spot of making a delightful hedge up to 3 feet. In the favoured south-west has grown to over 4 feet. A good fuchsia.

'**Drame**' (Lemoine 1880). Semi-double. Seedling of the well-known F. Riccartonii, it has inherited the hardiness of its parent and much of its vigour. The flowers, which are very free, have scarlet sepals and a violet-purple corolla, and they show up well against the yellow-green foliage.

'Dr. Foster' (Lemoine 1899). Single. This variety has perhaps the largest flowers of the recognized hardy varieties, being larger than many of the so-called 'greenhouse' types. The tube and sepals are scarlet and the corolla violet. It will not make a large hedge but an extremely colourful one.

'Heritage' (Lemoine 1902). Semi-double. Tube and sepals scarlet, corolla rich purple. Officially described as semi-hardy but listed here after seeing a fine 3–4 foot hedge of this variety at Chatham, Kent, after the severe winter of 1962–3. Flowers early and retains its flowers until the frost. May be cut down in severe cold but shoots strongly from the base. Well worth experimenting with.

'Madame Cornelissen' (Cornelissen 1860). Single. This is the only really hardy variety with a flower having a good size and having a white corolla. The tube and sepals are of rich scarlet. It makes an ideal shrub or low hedge to 4 feet. In mild coastal areas it can be grown even taller. To get the best from this variety it really does need the protection of a windbreak and to get extra height the base of the plants should be protected so that severe frost does not cut them down to ground-level. Good hedges have been grown with this variety.

'Margaret' (Wood 1937). Semi-double. Tube and sepals scarlet, corolla violet but base of petals tinged with white. Corolla has red veins. One of the best hardies raised by that specialist W. P. Wood. In mild districts it has made hedges up to 5 feet in height. Even in less favoured districts, given some shelter it will make a fine hedge. Well worth growing.

'Margaret Brown' (Wood 1949). Single. Tube and sepals rose-pink, corolla light rose-bengal. A distinct colour break in the hardy varieties. Near Dorking, Surrey, there is a fine hedge of over 3 feet high which survived the notorious winter of 1962–3. The flowers are small but produced in profusion. Like all strong-growing varieties it likes a deep cool root run. A variety that should be more widely grown.

'Mrs. Popple' (Elliot 1899). Single. The larger-than-usual flowers have scarlet tubes and sepals and corollas of deep violet-purple. A good hardy variety which, given a position it enjoys, will grow with vigour and make a handsome shrub. It has an erect habit and its upright branches carry a profusion of flowers.

'Nicola Jane' (Dawson 1959). Double. Tube and sepals are light scarlet. The whitish corolla is faintly veined throughout with pink. Growth is strong and upright. The flowers are completely weatherproof. One of the really outstanding hardy introductions of recent years.

'Phyllis' (Origin unknown). Semi-double. The sepals are waxy rose, the corolla rosy cerise. Discovered and introduced to commerce by H. A. Brown (1938), this variety was listed as half-hardy. The years have proved that in all but the most unfavourable districts this variety is very hardy. It is a grower of some vigour and although it may be cut down to ground-level by frost it is quite capable of sending up shoots 4–5 feet long the next season. It is quite exceptional in the amount of bloom it will produce. This variety should be tried by all growers.

'Prodigy' (Lemoine 1887). Semi-double. Catalogued under this name and that of 'L'enfante prodigue'. Cerise tube and sepals, corolla royal purple. An upright, bushy shrub which bears its medium-sized flowers very freely. Will make a delightful hedge of 3 feet high even in the most unfavourable places. Very hardy.

'Riccartonii' (Young 1833). Single. Tube and sepals scarlet, corolla dark purple. This is undoubtedly the greatest hedging variety ever raised. Massive hedges line the roads of Eire, Devon and West Scotland, and everybody who sees them marvels at their size and profusion of flower. It is possible to have a hedge of this variety practically anywhere in these islands provided that steps are taken to check the very coldest of east winds. It must be the ambition of every British hybridist to produce different-coloured fuchsias with the same vigour and constitution as this variety.

'Schneewitchen' (Klein 1878). Single. The tube and sepals are deep pink and they set off well the purplish-pink corolla.

The hardy flowers are small but most profuse and are borne on an upright, vigorous medium-sized shrub. A delight in any garden, it will make a very colourful hedge with the myriads of flowers well set off against the medium green foliage. Will bloom from July until the first frost. A reintroduction after having been 'lost' for many years.

'Thompsonii' (Thompson 1837). Single. Bright scarlet tube and sepals, with purple corolla. The slender, graceful, upright growth of this shrub bears a multitude of bloom. The flowers are smallish but are longer than its parent *F. magellanica*. Makes a handsome, open-looking hedge which is suitable for all but the most formal garden.

'Tresco' (Raiser and date unknown). Single. Discovered growing in the beautiful Tresco Abbey Gardens, Scilly Isles, and named accordingly. The flowers are small, with red tube and sepals and a purple corolla, and very freely produced. It has proved extremely hardy and grows with a vigorous spreading habit. If used for hedging, allowance should be made for this habit or the beauty of the shrub is lost.

CHAPTER XIV

Varieties Suitable for the Greenhouse

'Abbé Farges' (Lemoine 1901). Semi-double. Tube and sepals light cerise, corolla rosy lilac. The flowers are small, but extremely profuse. Rather slow to get going in the spring but soon catches up with the earlier starters. Does not like being moved about too much, as its branches are very brittle. A very charming and popular variety. Best grown as a bush.

'Abundance' (Todd 1870). Single. Tube and sepals scarlet-cerise, and corolla purplish magenta. A vigorous variety, always certain to grow and flower well. Flowers are of medium size and very free. Can be planted out permanently in some parts of the country. Can be grown as bush, half standard and standard.

'Alice Travis' (Travis 1956). Double. Tube and sepals carmine-cerise, corolla deep violet-blue. The sepals are of good substance, and the corolla is of good length. A large flower very freely produced. Needs training early to get a well-shaped plant.

'Amethyst' (Tiret 1941). Double. Tube and sepals pale red, corolla bluish violet. A vigorous upright variety bearing its flowers freely. An easy variety which seems to be a little unpopular because of its rather 'washed out' colourings. Can be grown as bush or standard.

'Angel's Flight' (Martin 1957). Double. White tube and deep pink sepals. Corolla white with bright pink stamens. The long flowers are freely produced on an upright bushy plant. An extremely attractive variety which is well worth growing.

'Anita' (Niederholzer 1946). Semi-double. Tube and sepals dark turkey red, corolla violet, with splashes of red shades that blend together. Very large attractive blooms on a vigorous upright-growing bush. It is fairly free in flowering and can be brought on quite early in the season. Good also in standard form.

'Anna' (Reiter 1945). Semi-double. Tube and sepals carmine, corolla magenta, flushed with carmine. The flowers are large and quite free and stand out well against the lush green foliage. Inclined to be rather lax in its habit of growth, it should receive careful training to produce a strong upright growth. Suitable for basket growing.

'Annie Earle' (Lye 1887). Single. Tube waxy cream, sepals cream with green tip, corolla bright scarlet. The medium-sized flowers are very freely produced, on a strong, upright-growing bush. Blooms early in the season. It is an attractive variety which will also make a good standard.

'Aphrodite' (Colville 1964). Double. Tube and sepals are coral-pink setting off a long and full corolla of pure white. American introductions in this colour range have all had a trailing habit, but this one is a good upright plant. The flowers are fairly large and are freely produced.

'Atomic Glow' (Machado 1963). Double. The tube is coral-pink leading down to the more glowing coral-pink sepals which are tipped green, with a deeper pink underside. The compact corolla is coloured cerise and the petals are splashed with orange at the base. Rather lax in growth but is self-branching, and lends itself well to training either as a bush, standard, or for basket decoration. An early bloomer and quite free flowering.

'Aunt Juliana' (Hansen 1950). Double. Tube and sepals carmine, corolla pale violet. The flowers are very large and fairly free. Growth is not strong, and is often not capable of supporting the heavy blooms. However, a good bush plant can be grown if time is given to tying in the early stages of growth. Also suitable for growing as espalier.

'Aurora Superba' (Raiser unknown). Single. Tube and sepals pale apricot, corolla deep orange-salmon. The medium-sized flowers are very beautiful and have a glowing effect. Best grown with the aid of heat, as in cool conditions the leaves curl and growing may become difficult. This is a variety which may disappoint the beginner, but will delight the grower with experience.

'Australia Fair' (Rawlins 1954). Double. Tube and sepals bright red, corolla white shaded and veined carmine. The flower is large and very freely produced. By habit it is an upright bushy shrub. Quite easy to grow and worth growing.

'Autumnale' (Metior). Single. Tube and sepals red, corolla purple. Flowers are small and rather late, and are secondary to the beautiful copper-red foliage and the unusual habit of growth which is distinctly horizontal. Deservedly popular with the enthusiast, and a handsome basket variety.

'Avalanche' (Henderson 1869). Double. Tube and sepals scarlet, corolla rich purple-violet, and shaded carmine at base. Flowers of medium size and very attractive. The leaves are a yellowish green, and form a pleasing background for the flowers. Forms a natural bush but can be grown as a half standard. Enjoys the shade. Well worth growing.

'Avalanche' (Schnabel 1954). Double. Tube and sepals are white, corolla creamy white, the flowers large with heavy broad sepals. Requires good cultivation to flower freely. The flower itself is extremely attractive, but tends to mark easily if grown in conditions that are too damp. It has an upright bushy habit of growth. Suitable also for growing as standard and half standard.

'Aviator' (Diener). Single. The tube and sepals are scarlet, corolla white, veined red. Very aptly named as the particularly long sepals are partially twisted like an aeroplane propeller. A beautiful variety admired by everybody.

'Avocet' (Travis 1958). Single. The short tube and narrow sepals are deep red, the barrel-tubed, single corolla is white.

As the stamens are concealed within the tube, the flowers have a unique appearance. The flowers are freely produced on a vigorous upright bush. Can also be grown as a standard.

'Aztec' (Evans and Reeves 1937). Double. Tube and sepals rich deep red, the full double corolla is purple to violet, with splashes of bright red in small stripes. The Americans describe the colours as those seen in the work of Aztec artists, hence the name. The flowers are large and fairly free. The foliage has a natural reddish tinge. An extremely vigorous grower which makes it ideally suited to training as a greenhouse climber.

'Ballet Girl' (Veitch 1894). Double. Tube and sepals crimson, corolla white faintly veined red. The sepals are reflexed and the corolla full. The flower really looks like a ballet dancer. This variety has received so much publicity since its introduction that the layman has given every fuchsia of similar description the same name label. For this reason some of the fuchsias offered under this name are not true to name. A fine grower and well worth growing. Grow either as bush, standard or half standard.

'Basket Strawberry Festival' (Haag 1956). Double. Tube and sepals bright red, corolla bright rose-pink and very full. Makes a vigorous, upright bush which produces its flowers freely. Over-feeding may cause the sepals to split. Grown well this variety is delightful as a bush plant, but is easily trained to half and full standard.

'Beauty of Exeter' (Letheren). Single, but many flowers tend to semi-double. Tube and sepals light cerise, the large corolla is bright rosy salmon. A good variety producing its large flowers very freely. To get the greatest benefit from this variety it should be trained from the time growth starts. Well worth growing.

'Bernadette' (Schnabel 1950). Double. Tube and sepals rosy cerise, corolla deep blue, fading to powder blue. The medium-sized flowers are very freely produced, on an upright, free-branching bush. A handsome variety, well worth growing.

'Billy Green' (English 1966). Single. A new addition to the 'Triphylla hybrids'. A strong, upright-growing plant with olive-green foliage. The flowers are produced in profusion and are pinkish salmon in colour. A useful addition to this popular section.

'Bland's New Striped' (Bland 1872). Single. Tube and sepals cerise, the corolla being of a rich dark purple, each petal of which has a rose-coloured stripe, or line, down its centre, extending from base to apex of the petals. As the flowers mature, the sepals recurve. Fairly free flowering. Upright bush requiring little training. Very distinct.

'Bloomer Girl' (Waltz 1952). Double. Tube and sepals red, corolla palest pink. The flower is large and fairly free. The weakness of the growth, which seems hardly able to support the large blooms, means that the best effect can only be obtained by constant tying. For this reason many nurserymen have dropped this variety from their catalogue. Can be trained for basket growing.

'Blue and Gold' (Reedström 1956). Double. Tube and sepals pink, corolla mauvish blue. The flowers are large and extremely beautiful. The foliage is citrus yellow in colour, and it acts as a foil to the flowers. A very good American variety.

'Blue Gown' (Milne). Double. Tube and sepals cerise, corolla violet-blue. A fine, strong, upright variety producing an abundance of large flowers. Can be made to bloom early in the season. A very worthy variety. Well named. Good for growing as bush, but will also make a standard.

'Blue Lagoon' (Travis 1958). Double. The short tube and broad recurving sepals are bright red, the full corolla is a rich deep purple cast on blue. The flowers are quite free and very attractive. Lax growth which needs tying to make a good bush.

'Blue Pearl' (Martin 1955). Double. Tube and sepals are white with faint pink tinge, with green tips to the broad, arch-

177

ing sepals. Rosette-formed corolla is violet-blue which retains its colour well. The petals of the corolla open flat against the sepals as the blooms fade with age. Good, vigorous, upright grower.

'Blue Petticoats' (Evans and Reeves 1954). Double. A white tube with sepals that are also white but faintly blush on the underside. Corolla opens as silvery lilac-lavender and ages to a pleasing orchid-pink. Flowers are of medium size and fairly free. Growth inclined to be rather weak.

'Blue Waves' (Waltz 1954). Double. Tube and sepals scarlet-cerise, the flared semi-double corolla powder blue, changing to violet with age. Corolla also flecked with cerise. The flowers are large and quite free. Growth upright and naturally bushy. Very attractive.

'Blush-o'-Dawn' (Martin 1962). Double. White tube and white sepals with green tips. The very full corolla is a silvery grey-blue. The flowers are of medium size and very long lasting. They show up well against the medium green foliage. Growth bushy and upright.

'Bobby Boy' (Fuchsia Forest 1965). Double. Tube and sepals reddish rose. Corolla opens as bluish rose and ages to rose with touches of orange on the outer petals. Has the appearance of a small rose. Flowers, though small, are produced in abundance. A bush variety of great charm.

'Bobby Wingrove' (Wingrove 1966). Single. Turkey-red self with the sepals paling a little at the ends. Flowers are small but very freely produced. Growth is that of a low compact bush. Becoming deservedly popular.

'Bobolink' (Evans and Reeves 1953). Double. Tube and sepals flesh-pink, corolla rich bluish violet. Flowers are of medium size and extremely free. A vigorous spreading habit of growth. Makes an excellent bush plant. One of the best American varieties introduced to this country, and an extremely easy variety to grow.

'Bon Accorde' (Crousse). Single. Tube and sepals ivory white, corolla pale purple suffused white. Blooms most prolific and stand erect in a most unfuchsia-like manner. Bushy and very upright habit of growth. Easy to grow, and attracts a great deal of attention. Well worth growing.

'Bountiful' (Munkner 1963). Double. The tube is pink. The broad sepals are also pink but with green tips. The fully double globular corolla is milky white with pink veining at the base of the petals. Growth is strong and upright and it flowers freely over a long period. Should make a good standard.

'Brandts 500 Club' (Brand 1955). Single. Tube and sepals pale cerise, the corolla similar but shaded orange. Almost a self-colour. The flowers are large and the sepals are very long. Very free blooming on a vigorous upright bush. Can also be grown as a standard.

'Candlelight' (Waltz 1959). Double. The tube and broad upturned sepals are white, although a faint blushing can be seen on the underside of the sepals. The large corolla opens as rose with overlapping petals of dark purple, the whole fading to bright carmine-red with age. Flowers are produced in good numbers. Will make a neat upright bush which must attract attention.

'Candy Stripe' (Endicott 1965). Single. Tube and sepals are pink with deeper underside. The tube is short and the sepals curl attractively upwards. Corolla opens pink but turns to pale violet with age. Makes an attractive upright bush.

'Cardinal' (Evans and Reeves 1938). Single. An extremely vigorous variety most suitable for growing as a greenhouse climber. The flower is almost a self-colour, having dark red tube and sepals and a scarlet-red corolla. The medium-sized flowers are freely produced, and stand out well against the medium green leaves.

'Cardinal Farges' (Rawlins 1958). Semi-double. The tube and reflexed sepals are pale cerise, and the corolla is white with distinct cerise veining. The flower is small but produced in great numbers. This is a sport from 'Abbé Farges' which it resembles in every way with the exception of the flower colourings. Very attractive.

'Carillon' (Rozain-Boucharlat 1913). Semi-double. Tube and sepals scarlet-cerise, corolla dark plum, splashed with pink. A large flower, very free-growing, on an easily grown bush plant. In some conditions the flowers lose their semi-double appearance and grow as singles. A good variety despite its 'old-fashioned' appearance. Will also make a charming half standard.

'Carioca' (Schmidt 1951). Single. Tube and sepals rosy cerise with a distinct green tip. The corolla opens a definite purple but when it is fully opened it has changed to carmine, and it then forms a neat cup shape. Flowers prolific and of medium size. Growth is strong and upright. A good American variety.

'Caroline' (Miller 1967). Single. The tube and sepals are cream, flushed pink, while the petals of the corolla open as a pale campanula-pink but mature with age to a cyclamen-purple. Growth is upright and the flowers of medium size are produced most freely. Foliage is a rather light green which sets the flowers off well. Should make a good standard.

'Chandlerii' (Chandler 1839). Single. Tube and sepals creamy white, corolla orange-scarlet. The flowers are large and free, and are produced on an upright bush. It is a good variety which despite its age will still stand comparison with more modern varieties. Will also make a good standard.

'Chantilly' (Kennett 1962). Double. A reddish tube leads down to white sepals. The full double-spreading corolla is white but some of the petaloids are flushed with pink giving what is described as an 'overall lacy appearance', hence the, name. The foliage is light green and quite large. Growth is lax and it is not free branching, although it flowers quite well. A remarkable bloom but a really untidy grower.

'Charlie Girl' (American). Double. Tube and sepals are

rose madder. The flared corolla consists of petals which are rich violet with rose-shading at their base. Flowers are large and freely produced. Growth is upright but needs tying when the blooms appear, as they are rather too heavy for the branches. An attractive variety which should become very popular.

'Cheerio' (Kennett 1967). Double. The tube and sepals are white. The tight, compact corolla is deep pink. This is a small-flowered double, but it flowers well with a neat bushy habit which does not require much training.

'China Lantern' (American). Single. The tube is white suffused with carmine. Both the sepals and corolla are bright carmine rose. Flowers are of medium size but very free. Growth is upright. Would make an excellent standard.

'Circe' (Kennett 1965). Semi-double. Tube and sepals pale pink. The corolla consists of four light blue petals, that fade to pale lavender, with a surround of pink petaloids which spread open almost flat. Profusion of flowers on an upright shrub. A good fuchsia. Will make a fine standard.

'City of Portland' (Schnabel 1950). Semi-double. The tube and broad sepals are carmine, the corolla petunia-purple, marbled with light carmine. The flowers are large and free. Does not enjoy a too warm greenhouse, preferring a cool shady position. Can be made to flower early in the season, when it will have a good flush of bloom, and then continue to flower intermittently until the autumn.

'Clipper' (Lye 1897). Single. Tube and sepals red, corolla claret-magenta. The sepals recurve to perpendicular as the flowers mature. Tall, vigorous and reasonably hardy, this variety will make a fine bush plant, or a handsome standard. Floriferous.

'Coachman' (Bright). Single. Tube and sepals pale salmon, corolla orange-vermilion. A very floriferous variety having a lovely long flower, with colours which really seem to glow. A fairly vigorous habit of growth which really lends itself to bush growing. Will flower over an extremely long period. Well worth growing.

'Confetti' (Martin 1965). Double. Tube and short sepals are pink. The medium double corolla is bright violet-blue with shorter outer petals of red, pink, and white, and many petaloids. The blooms resemble the fat confetti balls so common in the U.S.A. Blooms profusely on an upright bushy plant.

'Constellation' (Schnabel 1957). Double. Tube and sepals and corolla all creamy white. The flowers are large and very free. Growth vigorous and upright. An outstanding American variety produced in the quest for the perfect all-white flower. This variety will enhance any fuchsia collection and will always be admired. If it has a fault, it is that shared by all white flowers, whether they be fuchsias or anything else; the flowers are easily marked by water. An excellent bush variety which can also be grown as standard or half standard.

'Coralle' (Bonstedt 1906). Single. Tube, sepals, corolla all rich orange-salmon. The flowers are long and borne in terminal clusters. An *F. triphylla* hybrid, it shows very strongly its relationship with that species. Very showy flowers show up well against the blue-green foliage. Has only one fault, its susceptibility to frost damage.

'Core'ngrato' (Blackwell 1967). Double. The tube is pale coral as also is the outside of the sepals. The underside of the sepals are frosty salmon-pink. Corolla opens a light burgundy-purple, but changes to a salmon-burgundy with salmon-pink splashes at the base of the petals. A vigorous and upright grower which blooms freely.

'Corpus Christi' (Walker and Jones 1953). Single. Tube and sepals dark rose, corolla rose, shading to violet. The corolla is long and the flower generally large and quite free. Has an upright bushy habit of growth.

'Cosmopolitan' (Fuchsia Forest-Castro 1960). Double. Tube is salmon as also is the wide crêped sepals. The large double corolla consists of flaring, white-veined, pink petals with coral-pink twisted petaloids jutting out at the base. Flowers are large and freely produced. Growth is upright but not capable of supporting the weight of bloom without a lot of tying.

'Costa Mesa' (Fuchsia-La Nursery 1952). Semi-double. Tube and sepals a warm glowing pink, single to semi-double corolla, rose-bengal. The sepals are long and recurving. Flowers medium to large. Upright bushy habit of growth with a hard brittle leaf. Good variety.

'Countess of Aberdeen' (Forbes 1888). Single. Tube and sepals a delicate pale pink, corolla white suffused with palest pink. Growth is bushy but rather stiff. Very floriferous although the flowers are small. A charming variety which can be a little temperamental. Can be trained to a charming half standard.

'Crinoline' (Reiter 1950). Double. Tube and sepals white with a distinct green tip, corolla rose-pink. The flower is large and extremely free, and of classic shape. Growth is vigorous, upright and bushy. A most accommodating variety in every way and very easy to grow. Really worth growing. Suitable also as standard or half standard.

'Crown Jewel' (Schmidt 1953). Double. Tube whitish green, sepals white, corolla rose-pink with a light green centre. The corolla besides its lovely colouring is very ruffled and pleated. The flower generally is large and quite free. Growth is upright but not always able to support the blooms without much tying. A very striking variety. In the greenhouse can also be grown as a trailer.

'Curly "Q" ' (Kennett 1961). Single. The corolla is of four rolled petals of violet-purple. Pale carmine sepals reflex perfectly into circle which lie against the tube. Foliage is small and of an unusual grey-green in colour. Growth is spreading and self-branching. An attractive novelty.

'Dark Eyes' (Erikson 1958). Double. The short tube and the short, broad, upturned sepals are a deep red. The corolla is of a beautiful deep violet-blue. The petals are curled and rolled to form a full corolla of distinct shape. Growth is upright and bushy and there is an ample quantity of bloom.

'David Lockyer' (Holmes 1968). Double. Tube and sepals

white, corolla bright red striped with white. Has unique habit of producing secondary skirt from centre of corolla extending below existing corolla length making blooms appear much larger.

'Diawillis' (Niederholzer 1947). Single. Tube and sepals rose-bengal, corolla pale Tyrian rose. Flowers are large and quite free. The pastel colourings are beautiful, but growth is modest, and requires good cultivation to make a good bush.

'Don Pedro' (Twrdy 1872). Single. Tube and sepals and corolla all crimson. Very floriferous with flowers of medium size. Will flower very early in the season. Growth is bushy and upright. Another old variety which has stood the test of time, and still grown by fuchsia lovers.

'Don Peralta' (Tiret 1950). Semi-double. The long tube and sepals are dark red, corolla rose madder. The flowers are large, very free, but the more flowers the variety produces the more single they become. A fast, upright-growing variety with good foliage. Worth growing. Can also be grown in standard form.

'Dorothy Louise' (Schnabel 1952). Double. Tube and sepals pink, corolla orchid-pink. A lovely flower but not very free. Growth is upright, but slender, and needs tying early to get a good shape.

'Dr. Jules Gallwey' (Reiter 1940). Double. The long tube and the broad sepals are red, the corolla creamy white, veined with rose madder at the base. The flowers are very large and fairly free. The stems are long and slender and need tying to support the heavy blooms. Growth is quite vigorous and in the second year quite a large plant can be produced. Can also be grown in standard form.

'Dr. Topinard' (Lemoine 1890). Single. Tube and sepals light rose, corolla white. The corolla is flared and open. Very free flowering and growth naturally upright and bushy. An old variety but still worth growing.

'Du Barry' (Tiret 1950). Double. The tube and reflexed sepals are light pink. The very double corolla has the inside petals of purple, and the outside petals heavily marbled or entirely pink. The very heavy blooms are quite free, and are inclined to weigh down the upright growth. An attractive variety which needs tying to keep tidy.

'Duchess of Albany' (Raiser unknown). Single. Tube is waxy white, sepals recurving and of whitish pink. The corolla is pink. The flowers are of medium size and very free. Growth is good, of upright bushy habit with attractive, clean-looking foliage. A good variety which can also be grown as a standard.

'Duke of Wellington' (Haag 1956). Double. Tube and sepals red, corolla bluish mauve. The flowers are very large with a very full corolla, and quite free. Growth is upright and bushy. Attractive, and a 'must' for those who like the large American doubles.

'Dutch Mill' (American 1966). Single. Pale rose bengal tube and sepals. Corolla pale violet shading paler at base of petals. Flower is a perfect bell shape, and from a distance looks like numerous Dutch mills. Upright bushy grower.

'Earl of Beaconsfield' (Laing 1878). Single. Tube and sepals light carmine, corolla vermilion. The tube is long and of good substance, and the whole flower is approximately $4\frac{1}{2}$ inches long. Has a spreading habit of growth which is ideal for bush-grown plants. Extremely floriferous, and altogether a worthy variety.

'Easter Bonnet' (Waltz 1955). Double. Tube and sepals pink, flushed rose with green tips. The corolla is a dusky rose-pink, being of deeper colour at the base of the petals. The flowers are large and fairly free. It grows naturally to a medium-sized bush. A charming variety.

'Eileen Raffill' (Raffill 1944). Single. Tube rose, sepals white, flushed rose. Corolla Tyrian purple, rose red at base. The flowers are of medium size and very free. A few ties at the early stages of growth will make this variety grow into a neat bush. Quite unusual colourings.

'El Camino' (Lee 1955). Double. The sepals are broad, short and upturned, and coloured like the tube, a rose-red. Double corolla of large central petals and smaller spreading outer petals, white, heavily flushed and veined rose. Upright bushy grower with deep green foliage. Also makes a good standard.

'Elkhorn' (Brand 1949). Double. Tube and sepals carmine, corolla deep pink. The petals of the corolla are distinctly serrated, giving the outline of the horn of the elk. The flowers are large and fairly free, and the growth is that of a vigorous upright bush.

'Emma Calvé' (Lemoine 1912). Double. Tube and sepals cerise flushed carmine, corolla white. The sepals are carried horizontally over a perfect rosette-shaped corolla. The flowers are of medium size and quite free. Growth is upright and bushy.

'Enchanted' (Tiret 1948). Double. Tube and sepals rose red, corolla bluish mauve. The flowers are large and almost identical with 'Duke of Wellington' already described. When this variety was first introduced it caused quite a sensation on account of its size and amount of bloom. Unfortunately it did not have the growth to support its flowers, and it is rapidly being replaced by the 'Duke of Wellington'.

'Eppsii' (Epps 1840). Single. Tube of light pink, sepals bright cerise, lightly tipped green, corolla rosy magenta. The tube is rather long and the sepals are non-recurving. The foliage is light green. This variety is vigorous in growth and the flowers are moderately numerous. It makes a fine bush plant, but can be trained either as pyramid or standard.

'Eva Boerg' (Yorke 1943). Single to semi-double. Tube and sepals blush-white, corolla violet-purple, tinted rose. The flowers are of medium size and very free. Growth is of the low bush type.

'Fancy Pants' (Reedstrom 1961). Double. The tube and sepals are bright red. The corolla opens as light purple but ages to a beautiful reddish hue. A nicely shaped flower freely produced. An easy-to-grow variety suitable for bush, standard, or basket use.

'Fan Dancer' (Fuchsia Forest-Castro 1962). Semi-double. The tube and sepals are red and the corolla is a light orchid-blue with splashes of pink and red. The lay of the sepals make attractive rectangular-shaped blooms. Flowers are long lasting and very colourful. A worth-while variety.

'Fanfare' (Reiter 1941). Single. The long tube and the small pointed sepals are scarlet, and the small corolla is turkey-red. Receives its name from the flowers, which are the long horn shape. Very floriferous, the growth is strong and upright. A very distinct variety which will always attract attention. A first-class variety.

'Festival' (Schnabel 1948). Semi-double. Tube and sepals rose, corolla claret. The large flowers are freely produced on a vigorous upright shrub. The colouring gives a definite glowing effect. Worth growing. Grow also as standard or half standard.

'Firelite' (Waltz 1965). Double. Tube and sepals white. The sepals are distinctly pointed and curl on the edges. The rather 'fluffy', flaring corolla is a glowing carnival-red which retains its colour well, and does not fade noticeably as it ages as so many of this colour range do. Foliage is light green and acts as a good foil to the flowers. Growth needs training to get the best effect.

'Flirtation Waltz' (Waltz 1962). Double. Tube and sepals creamy white, flushed pink, corolla shell pink, deepening with age. Extremely free flowering, with medium-sized flowers of great substance. Growth is vigorous and upright, and it requires very little training to make it a well-shaped bush plant. A first-class variety.

'Floradora' (Waltz 1952). Double. Tube and sepals rose-pink, corolla deep violet-purple, with the petals touched pink. The medium-sized flowers are very free. Growth is upright and bushy. If not pinched out too much this variety will come into bloom very early in the season. It is a charming variety and well worth growing.

'Florentina' (Tiret 1960). Double. Tube and sepals frosty white. Corolla a distinctive smoky burgundy-red. Flowers large and free, but growth is rather weak. An unusual colouring.

'Flying Cloud' (Reiter 1949). Double. A very pale creamy white self, with a touch of pink at the junction of the tube and sepals. The medium-sized flowers are very free, and of delightful shape. A first-class variety but one which has a tendency to grow outwards instead of upwards. This variety is the yardstick against which all whites and near whites are judged. Should be in every collection.

'Formosa Elegans' (Story 1850). Single. Tube and sepals pink, corolla orange-cerise. The flowers are of medium size and freely produced. Growth is upright at first, but then tends to bend over. A beautiful variety, but it is inclined to be a little temperamental. A variety for the connoisseur.

'Formossima' (Girling 1846). Single. Tube and sepals are white, slightly suffused pink, the corolla being soft rose-pink. Very floriferous, this variety is quite distinct despite its age. The flowers are quite long and very beautiful. Growth is vigorous and upright, and besides making a fine bush plant it can be a handsome standard. Very easy and worth while.

'Fort Bragg' (Waltz 1957). Double. Tube and sepals rose, the sepals being broad and reflexed. The corolla is pale lavender tinged with blue. Petals slightly veined rose. The flowers are large and free, but the growth has to be tied at an early stage to make an upright bush. Can be grown as a trailer.

'Fuchsia Mike' (Conley). Semi-double. Tube and sepals bright red, corolla purple. The flowers are of medium size and fairly free. Growth is upright and bushy. Not an outstanding variety but still very pleasant.

'Gala' (Martin 1966). Double. Tube and sepals are salmon-pink although the ends of the sepals are tipped with green. The corolla is distinctive with blue petals in the centre surrounded by outer petals of both pink and lavender. The petals themselves have a slightly serrated edge. Growth is of a spreading, rather untidy, upright habit, but training can improve on this. Foliage is of medium green and rather on the small side for the size of the flower produced. Flowers are large and free.

'Gartenmeister Bonstedt' (Bonstedt 1905). Single. Tube and sepals deep flame, corolla orange-scarlet. This variety flowers profusely, bearing its flowers at the terminals in large trusses. The foliage is soft dark green with purple underlining. The habit is vigorous and upright. A good 'triphylla' hybrid.

'Gay Fandango' (Nelson 1951). Double. The tube and the long spreading sepals are carmine, the large double corolla is rosy claret. The petals of the corolla are in tiers, giving the flower extra length. The flowers are freely produced on a shrub which is upright and bushy. A popular variety and worthy of that popularity. Grows also in standard form.

'Gay Paree' (Tiret 1954). Double. The tube is white-flushed pale carmine, and the broad, upturned, green-tipped sepals are also white-flushed pale carmine on the outside, but the underside is a soft carmine-pink. Corolla is marbled purple, pink, and carmine. Flowers are large and very free and grow on a low bushy plant.

'Gay Senorita' (Schmidt 1939). Single. Tube and sepals rose-red, corolla dark lilac-rose. The corolla is a distinct bell shape. The large flowers are freely produced on a strong upright bush. Can be had in flower quite early in the season.

'General Wavell' (Whitcman 1942). Double. Tube and sepals pale cerise. The full corolla has outside petals of salmon colour, the inside petals being of deeper colour, shaded magenta. A vigorous and very floriferous fuchsia. A gay and attractive variety. Grows also in standard or half standard form.

'Genevieve' (Martin 1957). Double. Tube and sepals deep pink, corolla palish lavender. The flowers are of medium size with a very full corolla, and quite freely produced. Growth is a little weak, and much tying is needed to shape the plant well. Can be grown as a trailer.

'Georgana' (Tiret 1955). Double. Tube and sepals pink. Corolla has minor petals of pale lavender and outer ones of orchid-pink. A heavy bloomer with medium-sized flowers. Stiff upright grower with bright green leaves which are an excellent foil for the attractive flowers.

'George Travis' (Travis 1956). Double. Broad sepals, ivory at base, edged carmine, corolla lilac-blue, with silvery sheen. The sepals are not fully reflexed, and the corolla is full and globular in shape. Has a good upright habit of growth. A fine variety raised by one of Britain's leading raisers.

'Gertrude' (Warren 1952). Double. Tube and sepals pale rose, corolla smoky lavender, and petaloids marbled mauvish pink. The flowers are quite large and very free. Growth is upright and bushy. Very striking variety.

'Glendale' (Evans and Reeves 1936). Single. Tube, sepals and corolla, all coral-pink. Flowers produced in clusters, from early in the season until the fall. Growth is that of a vigorous spreading bush. The flowers are of medium size produced in great numbers. A very good, easy variety. Can also be grown in standard form.

'Glitters' (Erickson 1963). Single. Waxy-white tube and sepals, but the underside of the sepals is a lovely salmon. The spreading single corolla is a glowing orange-red. Medium-sized flowers are produced in profusion on a tall, vigorous, upright grower. An attractive easy plant to grow.

'Golden Dawn' (Haag 1951). Single. Tube and sepals light salmon, corolla delicate light shade of orange, changing to pale rose as the flower ages. The flowers are of medium size and very free. Growth is upright and bushy. Very pretty, and recommended.

'Golondrina' (Niederholzer 1941). Single. Tube and sepals rose madder, corolla magenta. The very long decorative buds open to form a large flower, with extra long reflexing sepals. Very prolific in flower. Growth is vigorous but rather spreading. Attention to training in the first year will produce a wonderful shrub the second year. A good variety, well worth growing. Suitable for growing also as a standard and will make a handsome basket.

'Gordon's China Rose' (Gordon 1953). Single to semi-double. Tube and sepals white at base shading to china rose, corolla magenta rose. Flowers are of medium size and quite freely produced. Apart from the colour, the flower is shaped very much like the more famous 'Lena'. Growth is upright and bushy.

'Governor Pat Brown' (Machedo 1962). Double. Tube and sepals which fold back in a graceful curve are flushed pink. Corolla violet with splashes of pink at the base of the petals. Flowers are large and very free and grow on an upright bush. Foliage is a yellowy green. Well worth growing.

'Green-n-Gold' (Rasmussen 1954). Single. This variety may well be a sport of 'Glendale', for the flower and habit are identical, and only the foliage appears to be different. The foliage is variegated green and gold.

'Grey Lady' (Reiter 1952). Double. Tube and sepals white, flushed pink, corolla pale grey-blue, but changing to lavender with age. The flowers are very large and are fairly free. Upright and bushy in growth, this variety is very showy and is worth growing.

'Guy Dauphine' (Rozain-Boucharlat 1913). Double. Tube and sepals carmine, corolla violet-plum. The corolla is made up of large petals, surrounded by a ring of smaller petals. The flowers are large and fairly free. Growth is upright and bushy. A good old variety.

'Hallowe'en' (Lagen 1938). Single. Tube and sepals creamy pink, corolla pink. Produces masses of medium-sized flowers. Growth is rather lax and it needs training to make a good bush. Given heat it will cascade naturally and make a good basket plant.

'Haphazard' (Hazard 1930). Double. Tube and sepals scarlet, corolla purple, with a white edge to the petals. Flowers are large and free. Growth is upright and bushy. This is one of the first varieties sent to Britain from the modern American raisers. Will remain a good variety for many years to come. Grow also as standard or half standard.

'Happy Talk' (Fuchsia Forest 1964). Double. Tube and sepals white faintly touched with pink. Corolla bright magenta and the petals flare into a star shape. Growth is rather lax and needs much tying to make a good bush. Attractive flower very freely produced.

'Haut Monde' (Hazard). Single. Tube and sepals creamy white, corolla dusky rose. Strong upright grower producing its medium-sized flowers very freely. Blooms early in the season and continues until the autumn. Good variety.

'Hayward' (Brand 1951). Double. Tube and sepals deep red, corolla violet-purple. The flowers are large and fairly abundant, and grow on a strong upright bush. A nice flower but lacks novelty. So many of the good old varieties are up to this standard and colour range.

'H. Dutterail' (Raiser unknown). Double. Tube and sepals scarlet, corolla plum-blue. The flowers are large and of very rich colouring. Growth is that of an upright bush. Not so free flowering as many other fuchsias, but very attractive.

'Heart Throb' (Hodges 1963). Double. The broad sepals are white with a pale carmine blush on the underside. The tube is white. Corolla is wide spreading, almost flat, consisting of large and smaller petals curled and folded, beautiful medium blue, white at centre and at base. As flower matures blue and rose petals are interspersed. A tall upright grower. A good variety.

'Heinrich Heinkel' (Rehnelt 1905). Single. Tube and sepals a light rose-pink, the latter being of deeper colour. Corolla the same colour as the sepals. Foliage is of dark olive green with deep magenta on the underside of the leaves. An *F. triphylla* hybrid with a vigorous habit, although in flower the growth is inclined to spread. The flowers are numerous and borne in clusters. A really first-class variety.

'Hidcote Beauty' (Webb 1949). Single. Tube and sepals waxy cream, flushed, and green tipped. Corolla clear salmon-pink. The medium-sized flowers are freely produced, and they show up well against the pale green foliage. Growth is upright and bushy. A gem which is worth a place in any collection of fuchsias. Will also make a charming standard.

'Hindu Belle' (Munkner 1959). Single. The tube is waxy white and the long, broad, upcurved sepals are flushed pale carmine. The long single corolla is of plum colour changing to red as flowers age. Flowers freely produced on an upright-growing plant.

'Hollywood Park' (Fairclo 1953). Semi-double. Tube and sepals cerise, corolla white, veined and flushed pink. The medium-sized bloom is produced in great numbers on an upright, bushy shrub. A very good fuchsia that is hardier than most American introductions. Will stand a lot of mishandling. Can also be grown in standard form.

'Ida' (Dixon 1869) Double. Tube and sepals scarlet, corolla rich deep purple flecked with mauve. The corolla is large and fluffy. The flowers are borne freely on a vigorous, upright bush. An old variety which has been rediscovered to the advantage of fuchsia enthusiasts. Will also make a good standard.

'Illusion' (Kennett 1966). Semi-double. The tube and sepals are white. The corolla has petals of pale lavender and pale blue shades with streaks of white running the length of each petal. The corolla is flared giving a more double appearance than there really is. Flowers are of medium size and fairly free. Growth is upright but needs a tie or two to maintain the bush shape.

'Impudence' (Schnabel 1957). Single. Tube and sepals pale cerise. Corolla white, veined rose. The four petals that form the corolla open out horizontally. The flowers are numerous, and of medium size. A very delightful novelty.

'Innocence' (Reiter 1952). Double. Tube and sepals pale rose, corolla clear white. Flowers are large and freely produced on an upright bush. Very easy to train, and flowers over a very long period.

'Iona' (Travis 1958). Single. Rather long tube, creamy wax, sepals pale rose. The large corolla is clear pink at the base, suffused lilac rose. The well-shaped flowers are produced on a strong upright bush. An easy variety to grow. One of the best British introductions of recent years.

'Iris Amer' (Amer 1966). Double. Tube and sepals white flushed pinkish orange. Corolla bright red with shades of carmine and splashes of orange. The medium-sized flowers are profuse on a short, jointed, upright shrub. An excellent introduction.

'Jack Shahan' (Tiret 1948). Single. Often incorrectly shown as 'Jack Sharon'. Tube and sepals rosy red, corolla rose bengal. The large single flowers are produced in large numbers, on growth that is inclined to be rather lax. With a few ties this variety will make quite a handsome bush. Will also make a fine weeping standard.

'Jamboree' (Reiter 1955). Double. The tube is a creamy white. Sepals creamy white on top, carmine red inside. Corolla carmine red. Flowers are very large and free and they take on the

Mission Bells

Morning Light

Mrs Lovell Swisher

Party Frock

Pat Meara

Pink Darling

Sleigh Bells

Snowcap

Stella Marina

Swanley Gem

Sweet Leilani

Swing Time

Thalia

Trase

White Spider

Yuletide

appearance of a carmine red self when the sepals reflex, and hide the tube. Growth is strong and upright, and the foliage is a very rich green. A very good variety. Can be grown also in standard form.

'Jeane' (Raiser unknown). Single. This variety is now catalogued under its synonym 'Genii'. Tube and sepals are pale cerise, the corolla is violet which ages to dark rose. The flowers are small but produced in great profusion. The small leaves are a pleasant yellowish green, and are a neat foil to the flowers. Very easy variety to grow and delightful in every way.

'Kaleidoscope' (Martin 1966). Double. Tube and sepals red, the very large double corolla is various shades of purple to pale lavender streaked with red and pink. The flowers are freely produced on a fairly vigorous upright bush. Well named as the colours of the corollas vary from flower to flower.

'Keepsake' (Kennett 1961). Double. Tube is pale carmine. Sepals are broad, short, and white on top, flushed pink beneath. Corolla is dianthus purple with four centre petals perfectly cupped and surrounded by many others of the same colour. Very free flowering with a medium-sized flower well formed and of some substance. Self-branching, upright, bushy type of growth supporting a light green foliage. A really charming variety which attracts attention.

'Keystone' (Haag 1945). Single. The tube to the base of the long sepals is the palest pink. The sepals fade from the pink to almost white, and end with a green tip. Corolla, a delicate apple-blossom pink. It is a prolific bloomer and a bush that is neat and upright. An excellent variety that will continually be admired. Will make a delightful half standard. Strongly recommended.

'La Bianca' (Tiret 1950). Single. Tube and sepals white, corolla white with the faintest flushing of pink on the petals. The flowers are quite free on a plant that is upright and bushy. Grown under too lush conditions in the greenhouse, the flowers are easily marked by any water that may settle on them. Put into the open the flowers become more resistant to damage, but the pink flushing in the corolla becomes more pronounced, and the tube and sepals also show pink. A charming variety that needs careful growing to get good results.

'Lace Petticoats' (Tiret 1952). Double. Tube and sepals pure white, corolla white, with the faintest tinge in the petals. The corolla is large and fluffy on flowers that are large, and very freely produced. Growth is upright, but inclined to straggle when the blooms are heavy on the branches. A very lovely flower but one which marks very easily. Not an easy variety to grow. and not easy to keep through the winter. One for the expert.

'La Fiesta' (Kennett 1962). Double. White tube and sepals. White of the sepals continues with the petals and petaloids of the double light dianthus purple, creating a new colour tone in fuchsias. The petaloids flare out well as flower ages. Flowers freely produced but growth is lax and a lot of tying is necessary to make a good plant.

'La Paloma' (Niederholzer 1940). Single. Tube and sepals are pale pink, corolla white, faintly veined pink. It produces many flowers of medium size, on a good upright bush. Very aptly named for the flowers can really look like doves about to settle. A charming variety. Can be grown to half standard.

'Lady Beth' (Martin 1958). Double. Thick crêpy-looking sepals are of pale rose. The giant double corolla is of the brightest violet-blue, with a phosphorescent sheen over it. It is a quite free-flowering variety, of an extremely strong upright habit. Makes a large bush, or can be trained as a climber.

'Lady Heytesbury' (Wheeler 1866). Single. Tube and sepals waxy white, corolla deep red to rose. A free-flowering variety with medium-sized flowers. It has a habit of growth which is neat and bushy. A very old variety but still an extremely good one.

'Lakeside' (Thornley 1968). Single. The tube is a bright scarlet-pink and the sepals are the same colour except that they have a very distinctive green tip. Corolla is almost blue when the flower first opens but as it ages the pink veining becomes prominent and the petals fade to lilac. Flowers are smallish but are produced in vast numbers. Habit is that of trailer, and with suitable training is suitable as bush, basket, espalier, or half-standard plant. An excellent introduction.

'L'Arlesienne' (Colville 1968). Semi-double. The tube and the long recurving sepals are pale pink. The long white petals of the corolla give a compact appearance. Upright growing bush type of plant which is quite free flowering.

'La Rosita' (Erickson-Lewis 1959). Double. Tube and recurved sepals are rose-pink. The corolla is orchid-pink. The flowers are of medium size and very free. Growth is upright. Grown in the sun the colours of the flowers become more strong and it will look a different variety to an indoor-grown plant. Suitable also as a standard.

'Lassie' (Travis 1959). Double. Tube and sepals bright red and the corolla is white. Flowers are large, full, and freely produced. Growth is rather lax in habit.

'Laura' (Niederholzer 1946). Single. The tube is ivory, the sepals neyron rose, and the corolla is fuchsia pink, with bright red edges to the petals. The beautiful long and large flowers are produced in quantity on a neat, bushy shrub. Requires a little training in the early stages of growth, but thereafter is easy and lovely.

'Lena Dalton' (Reimers 1953). Double. Tube and sepals pink, corolla blue, flushed rose. The large double blooms are freely produced on a compact small bush. Only recently introduced into Great Britain, but a variety which is quickly becoming popular.

'Libuse' (Lemoine 1879). Single. Tube and sepals are a waxy rose suffused cerise with the sepals tipped green. The corolla is scarlet-cerise. Flowers are of medium size and produced in fair numbers. An old-fashioned variety which seems to be losing favour.

'Lilac' (Haag and Son 1952). Single. Tube and sepals pale pink, corolla pale lilac. The flowers are large and free and keep their colour well throughout the time they are on the shrub. Growth is strong, but the shoots need tying in as they are inclined to spread. A good variety.

'Lilac Lustre' (Munkner 1961). Double. The short tube and broad upturned sepals are rose-red. The broad pleated petals of the corolla open powder blue but fade to lilac with age. The flowers are of medium size and are freely produced on an upright bushy plant bearing rich green foliage. A good fuchsia.

'Lilac-n-Rose' (Waltz 1958). Double. Tube and sepals rose flushed pink, corolla opens a clear pale lilac, but deepens to a clear rose-pink as the bloom ages. When the plant is in full bloom the colour effects are most attractive, with the blooms at different stages of development. The flowers are long and very freely produced. Growth is lax and much tying is necessary to get a good bush plant. Could be used for espalier work or for hanging baskets.

'Lollypop' (Walker and Jones 1950). Single. The tube and the upcurving sepals are iridescent pink. The large petals of the corolla open as a brilliant plum, and change to a deep peony purple as the flower develops. The flowers are huge and freely produced. Growth is weak and lax, and tying is necessary to get a plant worthy of carrying the wonderful flowers. Suitable also for hanging baskets.

'Lord Byron' (Lemoine). Single. The short tube and sepals are of deep cerise, the corolla a very dark violet. A distinctive fuchsia, the medium-sized flowers expand somewhat like a buttercup. Fairly floriferous with a habit that lends itself to growing in bush form.

'Lovable' (Erickson 1963). Double. Tube and sepals a quite deep red and the many petals of the large fluffy-looking corolla are orchid-pink veined with a deeper pink. A very beautiful and large bloom produced quite freely. Growth is rather lax, and the plant needs pinching and training from the early stages of growth. A flower which is always admired.

'Loveliness' (Lye 1869). Single. Tube and sepals creamy white, corolla rosy carmine. An extremely prolific bloomer, the medium-sized flowers grow on a vigorous upright bush. Very aptly named, this variety is easy to grow and can be recommended to all growers, expert or novice. Will also make a delightful standard.

'Luscious' (Martin 1960). Double. The tube and very wide sepals are dark red. Very large double corolla is dark wine and red marbled with orange. The large flowers are long lasting and fairly free. Growth is upright but willowy requiring support for the heavy blooms. Dark green foliage with clear red veins. A large beauty.

'Lustre' (Bull 1868). Single. Tube and sepals creamy white, corolla salmon-pink. The small flowers are borne in great profusion on short but sturdy branches. A variety that has always been popular because it is easy to grow, and can be had in flower over an extremely long period. An outstanding variety.

'Lye's Unique' (Lye 1886). Single. Tube and sepals waxy white, corolla salmon-orange. Flowers are smallish but very prolific. Growth is that of a sturdy upright bush. Although this variety has never been as popular as others by the same raiser, it is well worth growing.

'Madame Butterfly' (Colville 1965). Double. Tube and sepals turkey-red with very full double corolla of white, veined slightly with carmine. Growth is strong and upright and the flowers are very free. A variety which also makes an excellent standard.

'Major Barbara' (Schnabel-Paskesen 1958). Double. Tube and sepals pale rose, corolla pale violet-blue, ageing to old rose. Flowers are large and fairly free, but really too heavy for the rather lax habit of growth. A delightful flower, but a lot of tying is necessary to get a well-shaped bush. Suitable also for hanging baskets.

'Major Heaphy' (British). Single. Tube and sepals brick red, corolla scarlet. The flowers are small and borne in profusion on an extremely neat bushy plant. A very good variety having but one fault: in a dry atmosphere it will drop every bud and flower.

'Malibu' (Evans and Reeves 1953). Single. Tube and sepals coral-pink, corolla rose madder. The medium-sized flowers have a bell-shaped corolla. They are freely produced on a shrub which is very strong and upright. Needs pinching frequently to make it bushy.

'Mandarin' (Schnabel 1963). Semi-double. Tube and sepals are flesh colour. Petals of corolla are orange-carmine. Flowers large and very free. Foliage dark green and leathery. Growth is upright if grown in cool conditions but in the greenhouse it will cascade.

'Mardi' (European 1938). Double. Tube and sepals bright red, corolla light rose-pink. The flowers are large and grow in great numbers on a shrub which is bushy and upright. The colours and habit of growth are similar to the more famous 'Fascination', but the sepals remain horizontal and do not reflex as on the latter. Will also grow well in standard form.

'Mardi Gras' (Reedström 1958). Double. Tube and sepals red, corolla dark purple, with pink mottling on the petals. The flower is very large but not as free as one would like. Growth is that of a strong upright bush. A very charming variety when in full bloom, but the flowering period is rather short.

'Marietta' (Waltz 1958). Double. The medium-long tube and the broad upturned sepals are bright carmine. Large double corolla, a dark magenta-red with a few splashes of carmine on the outer petals, but as flower matures it turns a dark clear red. Strong upright growth makes this variety suitable for both bush and standard growing.

'Mary' (Bonstedt 1905). Single. A brilliant, self-coloured flower having tube, sepals and corolla of a vivid, bright scarlet. A 'triphylla' hybrid, it has 3-inch-long flowers with the distinctive shape of its parent *F. triphylla*. Foliage is a beautiful dark sage green, veined magenta, and heavily ribbed with the same colour. Habit is vigorous, and the flowers are borne in trusses at the terminals. A beautiful variety which is sometimes listed as 'Superbe'.

'Mary Lockyer' (Colville 1967). Double. The tube and broad sepals are red. The sepals are tipped green and have a crêped appearance on the underside. Corolla pale lilac marbled red. Strong upright grower.

'Mauve Poincaré' (Rawlins 1951). Single. Tube and sepals crimson, corolla clear mauve. This is a sport from the well-known 'Henri Poincaré' and has all the good attributes of that variety. The flowers are large and very free, and are very distinctive in shape, having a long corolla of classic bell shape. An early bloomer which stays in flower over a long period. Growth is upright and bushy. Also suited for growing in standard form.

'Mayfayre' (Colville 1967). Double. The very short tube and the broad sepals are red. The corolla is white with distinct carmine veining in the petals giving it a pink shading. The edges of the petals are rolled back making a most distinctive flower. Growth is upright and the flowers are quite freely produced.

'Mazda' (Reiter 1947). Single. Tube and sepals carmine, corolla orange-pink. Flowers are of medium size and grow on a vigorous upright bush. Very aptly named, for the blooms resemble Mazda lights as they are strung along the long branches. A lovely variety.

'Mel Newfield' (Schnabel 1952). Double. Tube and sepals red, corolla amethyst violet, marbled carmine. Flowers are large and fairly free. Growth is upright and bushy. A large American variety which will suit growers who enjoy large flowers of attractive colourings.

'Merle Hodges' (Hodges 1950). Double. Tube and sepals are rosy red, corolla soft powder blue, lightly veined and flushed pink. The sepals are broad and upturned, the corolla large and fluffy. Flowers are large and free. Needs tying up before the blooms appear as the growth is not really able to support the size and the number of blooms. A worthy variety if trained correctly. Can be used for hanging baskets.

'Merry Mary' (Fuchsia Forest 1965). Double. The tube is a pale pink, and the sepals where they meet the tube are also pale pink, but they fade away to almost white at their tips. The underneath of the sepals remains a deeper pink. The corolla consists of large white petals distinctly flushed by the pink veining. The outer petaloids are even more heavily splashed, with pink. An upright grower which needs support for the many large heavy flowers will weigh the branches down.

'Ming' (Jennings 1968.) Single. The tube and sepals are orange-red, flushed cerise, and the sepals have a green tip. The corolla is cherry-red. A very free-flowering cultivar which produces its wealth of bloom on a very upright growing shrub. One gets the impression that the famous cultivar 'Chang' is a parent of this variety. Should make a good standard.

'Minister Bouchier' (Lemoine 1898). Double. Tube and sepals red, corolla deep mauve. Flowers large and free. Growth upright and bushy. A variety that will stand a lot of mis-handling. Still listed in catalogues, but losing popularity to more modern introductions.

'Minuet' (Niederholzer 1943). Single. Tube and sepals red, corolla rich purple. Flowers medium and very free. The corolla opens quite flat. Growth is strong and upright. The flowers come very early in the season, and continue to bloom over a long period.

'Monsieur Joules' (Lemoine 1890). Single. Tube and sepals crimson, the sepals being curled right back to the tube. The corolla is violet-blue and the petals are smaller than the sepals. A very prolific bloomer with small dainty flowers. Growth is upright and bushy. An excellent variety, and a pleasure to grow.

'Monsieur Thibaut' (Lemoine 1898). Single. The tube and the long recurved sepals are cerise, the corolla magenta. The flowers are freely produced on a shrub that is extremely vigorous and upright. This variety will make a large bush extremely quickly. Also excellent as a standard.

'Monte Rosa' (Colville 1966). Double. The tube and up-turned curling sepals are deep pink. Corolla white with pink veining. Flower is large and freely produced. Growth is upright and shows the blooms to good advantage. Well grown it should look well on the show bench.

'Mrs. Churchill' (Garson). Single. Tube and sepals cherry-red, corolla pinkish white. The flowers are quite large and fairly free. Growth is upright and bushy. A nice variety which has not really made the grade as there are others of similar colour which are more free flowering.

'Muriel' (British, date unknown). Semi-double. The long tube is scarlet, with sepals of the same colour. The large corolla is bluish magenta. A very strong-growing variety which produces many flowers. It has light green foliage. A first-class all-round variety suitable for growing as bush, climber, basket, pyramid or standard. An easy variety to grow.

'Muriel Evans' (Evans and Reeves 1934). Single. Tube, sepals and corolla all red. The medium-sized flowers are very free and show up well against the light green foliage. Growth is strong and upright and the flowers will appear quite early in the season. Suitable for growing to standard form.

'New Horizon' (Reiter 1950). Double. Tube and sepals pale rose, corolla very light blue. The flowers are large and free. Growth is upright and bushy. Should be grown in a cool, well-shaded corner of the greenhouse to get the effect of the lovely pastel colours. Too much light hardens the colours, and a lot of the charm of this fuchsia is lost.

'Nicola' (Daglish 1964). Single. Tube and sepals are rich scarlet. Corolla violet-blue and opening flat like a saucer. Flower is quite large and very free. Growth is upright and bushy. The shape of the flower is distinctive, being a larger version of the well-known old favourite 'Swanley Gem', of which it is a seedling. Makes a truly wonderful half standard.

'Normandy Bell' (Martin 1961). Single. The tube and sepals are a pinkish white. The corolla is a single, large bell shape of light orchid-blue which matures to a good orchid-blue. The blooms hold their bell shape and are long lasting. Growth is upright and bushy. Good light green foliage. Excellent in every way.

'Not So Big' (Machado 1962). Single. Tube and sepals are a very pale pink. The corolla is blue which fades to a bluish mauve. The small flowers are very free and stand out on the stiff stems well above the foliage. Very upright form of growth which requires an extra pinch or two to get the best effect.

'Old Lace' (Brown and Soules 1953). Semi-double. Tube rose-red with sepals shading to rose-pink. The corolla is long, and is deep orchid blue, with pale blue edgings to the petals. Flowers are large and fairly free. Growth is strong and upright. A very attractive variety that wins a lot of attention. The corolla really does give an appearance of coloured old lace.

'Olympia' (Rozain-Boucharlat 1931). Single. Tube and sepals pale rose-pink, corolla carmine with crimson sheen. The medium-sized flowers show up well against the rather large foliage. Very prolific bloomer. Growth is vigorous and upright and needs an early pinch to develop a well-shaped bush. Excellent.

'Opalescent' (Fuchsia-La Nursery 1951). Double. Tube and sepals china rose, corolla pale violet, blended with opal. The flower is very large and for its size very free. Growth is willowy and needs plenty of support. A beautiful flower which has unfortunately poor supporting growth. Also suitable for hanging baskets.

'Orange Drops' (Martin 1963). Single. The tube and sepals are pale orange. The medium-sized corolla is orange. The blooms tend to hang in clusters. Growth is very upright. One of the most orange blooms to date.

'Pacific Grove' (Niederholzer 1947). Double. Tube and sepals crimson, corolla pale violet, with layer of petaloids pale crimson. Flowers are large but not so free. Growth is vigorous and upright. Would make a grand variety for exhibition purposes if only more flower was forthcoming.

'Painted Lady' (Hazard). Single to semi-double. Tube and sepals white, flushed pink, the flaring corolla is pink, shaded purple. Flowers are of medium size and very free. Growth is upright and bushy. A very lovely variety.

'Pantaloons' (Fuchsia Forest 1966). Semi-double. The tube is pink and the sepals light red. The corolla is described as a plum-burgundy but the most distinctive feature of the whole flower is that a number of the stamens are fasciated and have become modified petals hanging well below the corolla. It is an easy growing and free-flowering cultivar, but is lax in habit and therefore does need support if grown in bush form. Should make a nice hanging basket.

'Papoose' (Reedstrom 1960). Semi-double. The tube and sepals are bright red and the corolla is the deepest purple. A very small flower but the numbers produced more than make up for the lack of size. The plant is a wealth of colour throughout the whole season. Very versatile fuchsia suitable for all the known trained shapes.

'Pasteur' (Lemoine 1893). Double. Tube and sepals cerise, corolla very double, of rosette form, white veined cerise. Flowers are large and free on a bush which is upright and bushy. A very easy fuchsia to grow and still worth growing despite its age. Can be trained to standard form.

'Pat Meara' (Miller 1962). Single. The bell-shaped flowers have dark pink tubes and sepals. The corolla has veronica-blue petals which shade to white at the base. A seedling of 'Citation', it has the same wiry stems and upright growth, and although the flower opens in similar fashion it will not maintain the tidiness of its parent. Very floriferous. Will make an excellent bush.

'Peloria' (Martin 1961). Double. The tube and large sepals are a rich dark red. The large corolla is purple in the centre with red petals on the outer side. The corolla is distinctly star-shaped. Growth is upright and the flowers are fairly free. Has that something slightly different.

'Penelope' (Lemoine 1857). Single. Tube white suffused pink, sepals white slightly tipped yellow, corolla pale purplish pink. A very prolific bloomer, the small flowers growing on a shrub which is vigorous and upright. A beautiful fuchsia which, in its class, is still better than most modern varieties.

'Pink Cloud' (Waltz 1955). Single. Short tube and long, wide, upturned pink sepals curl slightly at tips. Corolla has four overlapping petals of clear pink which flare at maturity. A very free bloomer with dark green foliage. One of the largest single pinks. Very versatile. Can be grown as bush, standard, espalier, or pyramid.

'Pink Darling' (Machado 1966). Single. Tube and sepals flushed pink, corolla lavender-pink. Flower small, most profuse with the blooms standing out from the foliage. Growth self-branching, upright, with long stiff stems. A very charming variety becoming deservedly popular.

'Pink Dessert' (Knechler 1963). Single. The tube and the long sepals which grow straight out are pink. The corolla is a slightly paler shade of pink. The flowers are long and freely produced. Foliage is bright green. Growth is upright.

'Pink Flamingo' (Castro 1961). Semi-double. Tube and sepals are deep pink. Corolla a very pale pink but veined with dark pink. Flowers are large, long, and free. The sepals curl into various shapes, no two alike, giving the distant effect of a flock of flamingos. Growth is lax and for bush work needs much tying. Ideal as a weeping standard.

'Pink Galore' (Fuchsia-La Nursery 1958). Double. Tube and sepals deep pink, corolla candy pink. Flowers are large and free and stand out well against the dark glossy leaves. Growth is lax and much tying is necessary to support the weight of bloom. A beautiful flower let down by its habit of growth. Can be grown for hanging baskets.

'Prelude' (Kennett 1958). Double. The tube is red and the tapering sepals are white, sometimes flushed purple inside. The four centre petals of the corolla are royal purple and they are surrounded by pure white petals. When the bud opens only the white petals are visible, but as maturity is reached so the centre petals are exposed. Plant is upright and spreading. A beautiful flower.

'Pride of Orion' (Veitch). Double. Tube and sepals dark crimson, corolla white, veined cerise. Flowers are large and fairly free. Growth upright and bushy. A very old variety little seen these days. The bloom is lovely and it is to be hoped that at least some enthusiasts will see that this variety will not disappear altogether.

'Queen of Bath' (Colville 1966). Double. Tube and sepals pink. The corolla is a brighter pink. The very large flower is perfectly formed with sepals carried very high to give good depth to the bloom. Good upright habit of growth. Very free flowering for such a large bloom. Should also be tried as a standard.

'Raggardy Ann' (Brand 1952). Double. Tube and sepals cerise, corolla has a blue circle of petals, surrounded by an outer circle of rose-coloured petals. The flowers are large and quite free. Growth is vigorous and upright. This variety is very aptly named for the corolla is very ragged. A fuchsia with a charm of its own. Well worth growing for it is always admired. Can be trained to basket form.

'Rambling Rose' (Tiret 1959). Double. The tube, sepals, and corolla are all a soft pink. The medium-sized blooms are produced in great profusion on a lax-growing shrub. The amount of colour when in full bloom always causes favourable comment.

'Raspberry' (Tiret 1959). Double. The tube and the long pointed sepals are a pinkish white. The very full corolla is a deep raspberry-pink. The large flowers are produced on an upright bushy plant. Becoming very popular.

'Red Ribbons' (Martin 1961). Double. Tube and the curling sepals are a rich red. The corolla is white. The medium-sized flower is almost a smaller version of the more famous variety 'Texas Longhorn'. Growth is of the low bush type and the blooms are very freely produced. Also suitable for hanging baskets.

'Red Wings' (Tiret 1949). Semi-double. Tube and sepals red, corolla purple. Flowers large and very free. Growth upright but not strong enough to support the wealth of bloom. Needs several ties to get a good bush. Buds sometimes refuse to open when situation is too damp. Suitable also for basket growing.

'Regal' (U.S.A.). Single. Tube, sepals and corolla all rose madder. Flowers large and free. This is probably the most vigorous hybrid fuchsia ever produced. Will make a climber in the first year. Even then it is too strong for the smaller greenhouses unless the roots are severely restricted. A first-rate climber and nothing else.

'Regal Robe' (Erikson-Lewis 1959). Double. Tube and upturned sepals are a deep bright red. Corolla deepest royal violet-purple with many of the petals scalloped, giving an unusual appearance. Prolific bloomer on a tall, growing, upright bush. Should also be tried as a standard.

'Ridestar' (Blackwell 1965). Double. Tube and sepals red. Corolla deep violet-blue. A good double with an excellent upright habit of growth. Very floriferous. Excellent show fuchsia.

'Rose Bradwardine' (Colville 1965). Double. Tube and sepals deep rose-pink. Corolla large and fluffy and is lavender splashed with orchid-pink. The sepals curl right back. Very robust upright habit. Excellent introduction.

'Rose of Castile' (Banks 1869). Single. The waxy tube and sepals are white, the corolla purple flushed with rose. The flowers are small but produced in abundance. Growth is upright and very bushy. Foliage light green. An old variety, but one that will always be grown. Easy to grow and accommodating in every way. For those who love the small-flowered fuchsias this variety is a 'must'.

'Rose Pillar' (Reiter 1940). Single. Tube and sepals rose with a green tip to the sepals. Corolla clear neyron rose. Almost a self-coloured flower. The corolla is open very much in the same manner as the well-known variety 'Display'. Growth is vigorous and upright, and the foliage large. Will make a large bush, or can be trained to make an attractive climber.

'Royal Flush' (Hodges 1952). Single. The tube and the broad, long, upturned sepals are deep red. The corolla, which opens back until it is flat, has round petals which are royal purple, flesh colouring at base, with heavy veining at the centre. The whole flower is large and quite free. An eye-catching variety well worth growing.

'Ruthie' (Brand 1951). Double. The tube is white streaked with red. The large crêpe sepals are white as they open, but as the flower ages the red spreads from the tube to flush the whole of the sepals. The medium corolla opens violet-purple with pink petaloids but turns more reddy as the flower matures. A most floriferous variety with an upright habit of growth. Becoming more popular as it becomes known.

'Sacramento' (Reiter 1941). Single. Tube, sepals and corolla, all light carmine. The sepals recurve gracefully, giving the medium-sized flower a classic fuchsia shape. Prolific bloomer with attractive foliage. Vigorous, good, upright habit. An excellent variety. Will also make a good, tall standard.

'Sahara' (Kennett 1966). Double. Tube and sepals pink, medium double corolla with centre opening dianthus purple with petal overlay of salmon-pink. The corolla is very compact. Medium-sized flowers very free on an upright growing plant. Could be grown as a standard.

'Santa Lucia' (Munkner 1956). Double. The tube and the broad spreading sepals are deep red. Corolla is orchid-pink with prominent rose veining on all petals. The outer petals are somewhat attached to the sepals, and as they open so the whole flower seems to flare. Strong upright grower with large flowers.

'Santa Maria' (Nelson 1952). Double. The tube and the spreading sepals are red, corolla white veined red, with a ring of very small red petals surrounding them. Flowers are large and produced in fair numbers. Growth is upright and bushy. A most attractive variety, very easy to grow, and should be more popular.

'Santa Monica' (Evans and Reeves 1935). Double. Tube and sepals red, corolla very light pink, shading to a darker pink at base of petals. Petals have red veining. Flowers are large and unfortunately not very freely produced. This variety has a reputation for being a late bloomer, so care must be taken not to 'pinch out' too much, or further growth will come at the expense of flowers.

'Sarong' (Kennett 1963). Double. The tube is white, and the long twisting sepals are white, but flushed with pink and tipped green. The large double corolla is violet-purple, marbled with a pink which becomes more noticeable as the bloom opens. The corolla flares wide at maturity. Growth is very vigorous but lax, training is needed at an early stage in growth. Flowers fairly freely but the flowers are inclined to be rather overwhelmed by the large medium green foliage.

'Satellite' (Kennett 1965). Single. Tube and sepals white. Large corolla of dark red that fades to bright red with streaks of pure white either on the outside or the inside of the petals. Flower quite large and very free. Growth upright.

'Satin's Folly' (Martin 1962). Single. Sepals maroon, very long and silky. Corolla reddish purple. The flower is very long and freely produced. Growth is a little lax but with careful tying will make an attractive bush.

'Seventeen' (Reiter 1947). Double. Tube, sepals and corolla all a soft, clear pink. Flowers are large and very freely produced. Growth is upright and bushy. When first introduced this variety was immediately popular because of the beautiful colour of the flowers. Unfortunately it has turned out to be a difficult plant to grow in this country, although the Americans say that it is a success in U.S.A. It is felt that new stock from America might restore the vigour that the Americans say is there.

'Shangri La' (Martin 1963). Double. Tube and sepals dark pink. Corolla pale pink with a heavy deeper veining. Opens up to a large unusual pink and full bloom. Growth is upright and the flowers are very free.

'Shasta' (Kennett 1964). Double. The tube is pink and the sepals are pink with white edges. Corolla white with touches of pink on the petaloids. Has serrated edges to the petals. Flower is large and very free. Growth is medium bushy.

'Snow Flurry' (Schnabel 1952). Semi-double. Tube and sepals are white, with a distinct pink flushing on the sepals. Corolla white, faintly tinged with pink at base of petals. Very profuse bloomer with medium-sized flowers. Growth is inclined to be rather lax and some tying up is needed. Very attractive variety, and always admired. Grows easily as a trailer.

'Sophisticated Lady' (Waltz 1964). Double. The tube and large wide sepals are a very pale pink. The short, full corolla is white. Flowers are free but growth is a little too willowy and yet not sufficiently strong for good basket-work.

'Southgate' (Walker and Jones 1951). Double. Tube and up-turned sepals are medium pink, corolla very soft powder pink. Blooms are large and produced in great numbers. Growth is vigorous and upright. One of the best varieties the Americans have sent us. It is easy to grow and requires hardly any training. Colour is beautiful and there is plenty of it. Every collection of fuchsias would be enhanced by this variety. Will also make a beautiful standard.

'Strawberry Festival' (Haag). Double. Tube and sepals bright red, corolla pale pink, veined carmine. Flower is large and very free. Growth is upright and bushy, but is inclined to be soft so that the weight of bloom becomes too much for the branches to bear. A good fuchsia but one that needs tying up. Suitable also for hanging baskets.

'Sunburst' (Reiter 1946). Semi-double. Tube and sepals pale carmine, corolla crimson. Medium-sized flowers grow freely on a shrub that is upright and bushy. When in full bloom this variety has a fiery appearance which is most attractive. Very aptly named.

'Sunkissed' (Tiret 1957). Double. Tube and sepals pale pink, with a tip of green to the sepals, corolla jasper red, spreading to crimson. The medium-sized flowers are freely produced and very lovely. It is a strong grower, easy, and a delight to grow. A variety that is always admired. Will also make a good standard.

'Sunrise' (Reiter 1942). Single. Tube and sepals white flushed rose, sepals tipped crimson, corolla clear scarlet. Flowers are of medium size, very free, and can be had quite early in the season. Growth is upright and bushy. A charming shrub.

'Swan Lake' (Travis 1957). Double. Short tube eau-de-nil, long broad sepals, white with pink flushing at base. The globular corolla is chalk white, with the faintest pink tinge in it. Flower is fairly large and fairly free. Growth is upright. The soft colourings are extremely pleasant. An excellent British intro-duction.

'Swanley Yellow' (Cannell 1900). Single. Tube and sepals orange-pink, corolla orange-vermilion. The tube is long. Flowers are freely produced on a plant that is vigorous and upright. Foliage is large and has a distinct bronze cast. The name is misleading as nobody can describe the flower as yellow. A good variety with unusual colourings.

'S'wonderful' (Castro 1961). Double. Tube and sepals are pink. Corolla consists of pale lavender inner petals and orchid-pink outer petals. The flowers are of medium size and very free. Suitable for bush and basket.

'Tahiti' (Schnabel 1963). Double. The tube and the sweeping sepals are of the palest pink. The huge fully double corolla has rose-lake colourings. The flowers are freely produced and show up well against the medium green foliage. Growth is strong and naturally bushy.

'Tammy' (Erikson 1962). Double. Tube waxy white. Long broad sepals a lovely pink, spreading, twisted, and upturned. Large outspreading corolla mauve-lavender, streaked rose-pink. Flowers very free. Growth vigorous and spreading. Suitable for bush and basket.

'Television' (Evans 1950). Double. The tube is white, the sepals are white on top, with pink underside. Corolla is deep orchid, splashed fuchsia pink at base. Growth is medium, upright and very bushy. Flower is large and very freely produced. A truly magnificent flower and quite unusual. Care must be taken when pinching out as shrub becomes too bushy and unwieldy, with subsequent loss of flower.

'Texas Longhorn' (Fuchsia-La Nursery 1961). Double. The tube and exceptionally long sepals are bright red. The very white petals of the corolla are lightly veined with red at their base. From buds of over 4 inches in length the sepals can spread to a span of over 7 inches. Growth is upright but is rather poor and quite unable to support the large blooms without much tying. Not easy to grow but the flowers are most rewarding.

'The Doctor' (Castle Nurseries). Single. Tube and sepals pale rose, corolla rosy salmon. Very prolific bloomer, the flowers being of medium size. Growth is vigorous, upright and bushy. A wonderful variety to see in full bloom, and very easy to grow. Besides making a good, large bush, this variety is ideal for training as a standard.

'Three Cheers' (Creer 1968). Single. The tube and sepals are bright red. The four petals that comprise the corolla are violet-blue but each has a white marking where they join the rest of the flower. The corolla opens flat and one would imagine that the popular 'Swanley Gem' was a parent. Name is derived from the distinct sections of the bloom, red, white and blue. An upright, free-flowering, worthwhile cultivar.

'Ting-a-Ling' (Schnabel 1958). Single. Corolla, tube and sepals are all white. The petals flare in a similar fashion to the popular variety 'Display'. Growth is strong and upright and it flowers prolifically over an extremely long period. Suitable for growing in both bush and standard form. Must become very popular as more people get to know of it.

'Traudchen Bonstedt' (Bonstedt 1905). Single. Tube and sepals are light rose, the corolla light salmon. An *F. triphylla* hybrid, bearing 2¼-inch-long flowers of distinctive shape. Flowers are profuse and borne in terminal clusters. The foliage is a light sage green, with veins and ribs of a lighter shade. Growth is upright and bushy. A well-grown bush of this variety when in full flower is a wonderful sight. Really worth growing.

'Uncle Charley' (Tiret 1949). Semi-double. Tube and sepals rose-red, corolla lavender-blue, with lilac shading. Flower large and very freely produced. Growth is upright and bushy, and quite vigorous. An excellent variety which has become very popular. Suitable also for growing as a standard.

'Unique' (Haag 1950). Single. The tube and broad sepals are rose madder, the corolla also rose madder, but marbled at the base of the petals with imperial purple. Flowers are of medium size and very freely produced. It makes a dwarf but sturdy upright bush. An attractive variety which is easy to grow and will stand a lot of rough handling.

'Upward Look' (Gadsby 1968). Single. Tube and sepals are carmine but the sepals are tipped green. The corolla is pale rosine-purple. This cultivar is the result of a cross between the very popular 'Bon Accorde' and 'Athela' and it has 'Bon Accorde's habit in as much that the flowers stand erect instead of hanging in the recognized fuchsia manner. Almost as prolific in flower as its famous parent, and not so stiff in growth thereby giving a bushier looking plant. Makes a good and unusual standard.

'Valentine' (Reiter 1948). Semi-double. Tube and sepals white, flushed rose, corolla imperial purple, shading to white. Flower is large and quite free. Growth is lax and inclined to spread. The flower is extremely unusual but the growth needs much tying to make a well-shaped bush. Can also be used for hanging baskets.

'Vanessa' (Colville 1964). Double. The tube and long sepals which curl upwards are pale carmine. Corolla lavender-blue. The large flowers are very free and are produced on a strong upright bush.

'Ventura' (Evans and Reeves 1951). Single. Tube and sepals rose madder, corolla coral-pink. A prolific bloomer with medium-sized flowers borne in terminal clusters. Growth is vigorous, upright and bushy. A very distinct variety well worth growing for the amount of bloom it will give. Can also be trained as a standard.

'Vera Sergine' (Lemoine 1892). Semi-double. Tube and sepals white, flushed very faintly with a very pale pink, the sepals are edged in pink. Corolla also white flushed pink. Flowers are large—about 3 inches long—and freely produced. Growth is upright and bushy. Many of the beautiful 'whites' now listed have this hybrid somewhere in their ancestry. This is probably of more use to the hybridizer than the normal grower, for it is not so showy as many of the more modern varieties.

'Viva Ireland' (Ireland 1956). Single. Tube pale pink with the sepals a little deeper. Corolla clear lilac-blue, flecked soft pink. Flowers of medium size and very free. Growth is willowy. Grow as bush and it will also cascade for a basket.

'Voltaire' (Lemoine 1897). Single. The tube and the recurved sepals are scarlet-cerise, the expanding corolla rosy carmine. Flowers are of medium size and very free. Growth is sturdy and upright. Dismissed by many as old-fashioned, but a variety which is easy to grow, and one which blooms very early in the season. Continues in flower for a long time. Trouble-free and charming. An excellent bush variety which also trains well to standard and half standard. Quite hardy.

'War Paint' (Kennett 1960). Double. The short tube and broad flaring sepals are white. Corolla dianthus purple with coral-pink marbling, fading to reddish purple. Flowers large and free. Vigorous upright grower suitable for bush or standard.

'White Queen' (Doyle 1899). Single. The long tube and the sepals are cream, slightly tinged with pink, corolla orange-vermilion. Flowers are large and free, and very attractive. Growth is vigorous but rather lax in habit. Needs several ties to make a well-shaped bush. Suitable also for hanging baskets.

'Winston Churchill' (Garson 1942). Double. The tube and reflexed sepals are of rose-pink, the large double corolla is blue, shaded with magenta pink. Good foliage of dark green, with very strong stems, and strong upward growth which needs no support. A fast grower which blooms well throughout the season. It is inclined to be difficult to keep during the winter except in the most congenial conditions. Can also be trained as a standard.

'Yankee Doodle' (Hodges 1953). Single. The short tube and the upturned sepals are dark red, and the single, flat corolla is white, with very irregular elongated blotches and stripes of bluish purple. Practically every flower has different markings. The medium-sized flowers are very freely produced. Growth is upright and very bushy. Rather bizarre, but has an attraction all its own.

CHAPTER XV

Varieties Suitable for Summer Bedding

'**Army Nurse**' (Hodges 1947). Semi-double. Tube and sepals deep carmine, corolla violet-blue, flushed pink. Flowers are of medium size and very profuse. Growth is vigorous and upright. A variety well able to stand the rigours of an English summer without suffering damage. Well grown under greenhouse conditions it can be trained to both pyramid and standard. Is totally hardy in some parts of the country.

'**Brilliant**' (Bull 1865). Single. Tube and sepals scarlet, corolla violet magenta. The flowers are of medium size and very free. Growth is upright and very bushy. An old variety that is still worth growing. It will bloom in the open from very early in the season until the first frost. In the greenhouse will make a fine standard.

'**Charming**' (Lye 1895). Single. Tube and sepals carmine, corolla reddish purple. Flowers are medium to large and extremely free. Growth is neat and bushy. Foliage is a bright yellowish green. A variety that has received the Award of Merit of the R.H.S. Will bloom in the worst conditions that an English summer can give. Can be trained to standard or half standard under greenhouse conditions.

'**C. J. Howlett**' (Howlett). Single. Tube and sepals reddish pink, corolla bluish carmine. Flowers are small to medium in size and very freely produced. Growth is upright and bushy. A variety that, even in the open, will bloom early in the season, and continue to produce flower over an extremely long period.

'Cloth of Gold' (Stafford 1863). Single. Tube and sepals scarlet, corolla purple. The flowers are of medium size but there are very few of them. This is first and foremost a foliage plant and must be treated as such. The lovely golden foliage has a red underside to it. It is an ideal edging plant, on its own, or used in conjunction with other plants. Growth is medium and bushy.

'Conspicua' (Smith 1863). Single. Tube and sepals are the brightest scarlet, the corolla milky white. The flowers are smallish but most profuse. Growth is upright and bushy. This is a variety which really delights in being planted out during the summer months where it will bloom and grow more satisfactorily than if kept in the greenhouse. Can be had in bloom the whole summer.

'Duke of York' (Miller 1845). Single. Tube and sepals cerise, the latter recurving. Corolla a bluish violet. Although the flowers are of medium size they are most profuse. Foliage is a light green. Growth is strong, upright and bushy. Grows and flowers well in the open. In the greenhouse will make a fine standard.

'Elsa' (British). Semi-double. Tube and sepals are ivory pink. Corolla rosy purple. Flowers are medium to large and very freely produced. Foliage is a light green. Growth is inclined to sprawl but is quite vigorous. Very similar to the variety 'Lena', with which it is often confused. Hardy in favoured situations, and in other places will make a high-quality bedding plant. In the greenhouse will make a fine standard. Can be had in flower early in the season.

'Emile Zola' (Lemoine 1910). Single. The tube and the reflexed sepals are scarlet-cerise. The open corolla is rosy magenta. The medium-sized flowers are freely produced on shrub that is upright and bushy. A variety that has been given the Award of Merit of the R.H.S.

'Forget-Me-Not' (Niederholzer 1940). Single. Tube and sepals pale pink, corolla pale blue. The small to medium flowers are borne in great profusion. Growth is vigorous, upright and bushy. The mass of attractive flowers with their prominent long pistils makes this variety extremely popular.

'Golden Treasure' (Carter 1870). Single. Tube and sepals scarlet, corolla purplish magenta. Flowers are of medium size and have the reputation of being scarce. Given some care in cultivation it is surprising the amount of flower this variety will produce. Grown really for the beautiful foliage which is golden with red veining. Growth is strong but very spreading. Whole beds have been given over to this variety, which will make a golden mat with the red veining standing out quite distinctly.

'Hapsburgh' (Rozain-Boucharlat 1911). Single. Tube and sepals of waxy appearance and pale pink in colour. Corolla deep violet. Flowers are of medium size and extremely freely produced. Growth is upright and bushy. Bedding in the open has no effect on the amount of flower this variety will give.

'Henri Poincaré' (Lemoine 1905). Single. Tube and sepals red, corolla violet-blue. The sepals are reflexed, and the corolla long and bell-shaped. The flower is large and very free. Growth spreading and bushy. The shape of the flower is quite distinct. Inclined to grow soft in the greenhouse but is fine outside. A gross feeder that requires good cultivation.

'Heron' (European). Single. Tube and sepals scarlet, corolla bluish violet, veined cerise. The flowers are large and borne in great profusion. Growth is strong, upright and bushy. The flowers are of good substance and have good weather resistance. This variety is well known for the mass of bloom it will produce. Under greenhouse conditions will make a very stately standard.

'Kolding Perle' (Danish). Single. The long tube and sepals are waxy white, and the corolla is pink with salmon shades. The medium-sized flowers are produced in abundance on a vigorous upright bush. Its origin is unknown except that it grows well in the Botanical Gardens in Copenhagen. Now obtainable in Britain.

'Letty Lye' (Lye 1877). Single. Tube and sepals flesh pink, corolla carmine. Flowers are of medium size and borne freely. Growth is vigorous, upright and bushy. Not an easy variety to obtain but a fine one for summer bedding.

'Leverkusen' (Hartnaeur). Single. Mostly catalogued under the name 'Leverhulme'. Tube, sepals and corolla cerise. The tube is the longest part of the flower and it is also rather thick. The sepals and corolla are both small. Flowers are numerous and borne in clusters. Growth is upright and bushy. A triphylla hybrid which has a strong leaning toward the 'Magellanica' types. Excellent for bedding.

'Lye's Favourite' (Lye 1886). Single. Tube and sepals flesh pink, corolla orange-cerise. Very free blooming with flowers of medium size. Growth is sturdy, upright and bushy. The colourings are delightful and add distinction to a flower bed.

'Moonlight Sonata' (Blackwell 1964). Single. Tube and sepals bright pink. Corolla light purple, flushed and veined pink at base. Medium-sized flower produced in profusion. Growth is bushy. Grown well this variety is a mass of bloom for a very long period.

'Port Arthur' (Story 1869). Double. Tube and sepals are red, corolla bluish purple. A free-flowering variety having medium-sized blooms. Growth is sturdy and very bushy. This variety has been proved to be very hardy in some parts of the country. Ideal for bedding purposes.

'Ron Honour' (Vicarage Farm Nurseries 1969). Single. The tube and sepals are crimson and the corolla is purple, veined heavily with crimson. Each petal is rolled into a tube. Good, strong habit of growth which requires but little training to make an excellent bush plant. Very free flowering and seems to enjoy an outdoor position. Should also make an excellent exhibition plant if given that extra attention needed for that purpose.

'Sunray' (Milner 1872). Single. Tube and sepals bright cerise, corolla rosy carmine. The smallish flowers are of secondary consideration and are few and far between. The attractive foliage for which this variety is grown is light green, suffused cerise, the edges of the leaves being cream. Growth is upright and bushy.

'Thalia' (Turner 1855). Single. The long tube, tiny sepals and the small corolla are all a rich orange-scarlet. The long flowers are borne in profusion, growing in trusses at the terminals. The foliage is a dark olive green with distinct magenta veins in the centre, and along the ribs of the leaves. Of quite vigorous habit, it will quickly make a good bush. For over a hundred years this variety has been used extensively for summer bedding.

'Tolling Bell' (Turner 1964). Single. Tube and sepals scarlet, corolla white. The corolla is shaped like a church bell. Extremely free flowering and growth is strong and upright. Well worth growing. Can also be trained as a standard.

CHAPTER XVI

Varieties Suitable for Permanent Bedding

'**Alice Hoffman**' (Klese 1911). Single. Tube and sepals rose, corolla white. The flowers are small but borne in profusion. Growth upright and very bushy. The flowers show up well against the bronzy green leaves. A dwarf shrub ideal for the front of the shrub border, the window box, indoor culture and even for a permanent bed of its own. Quite hardy. Will make a charming table standard.

'**Blue Boy**' (Fry 1889). Single. Tube and sepals dark pink, corolla violet-blue. Flowers are small but are borne in profusion. Growth is strong and upright. In a comfortable sheltered position, say in the front of a shrubbery, this fuchsia will survive all but the most severe weather. Cut down in the winter it sends strong shoots from the base to flower from early July until the frost comes again. With greenhouse culture this variety will make a fine standard.

'**Carmen**' (Lemoine 1893). Semi-double. Tube and sepals cerise, corolla purple. Flowers are small but are produced in great numbers. Growth is dwarf and bushy. Looks charming on the rockery. Very hardy.

'**Chillerton Beauty**' (Bass 1847). Single. Has been catalogued under the name 'Query'. Tube and sepals pale rose-pink, with even paler edges to the sepals, corolla mauvish violet. Flowers are small but extremely free. Growth is upright and naturally bushy. An ideal shrub for a sheltered position. Has grown and flourished outdoors at Chillerton, Isle of Wight, for well over fifty years. In the right position will grow up to 3–4 feet.

'Cupid' (Wood 1946). Single. Tube and sepals cerise, corolla deep bluish pink. Heavy bloomer with flowers of medium size. Growth is low and bushy. Has proved fairly hardy in sheltered positions.

'Daniel Lambert' (Lee 1857). Single. Tube and sepals pink, corolla light magenta. The medium-sized flowers are borne freely. Growth is upright and bushy. In the south and south-west of Great Britain this variety has flourished outdoors for many years without suffering any ill effects. Worth trying.

'David' (Wood 1937). Tube and sepals scarlet, corolla rich purple. Flowers are very small, but this lack of size is compensated for by the number borne. Growth is short and upright. Very hardy and is ideal for the edge of the border. A good window-box variety.

'Dunrobbin Bedder' (Melville 1890). Single. Tube and sepals scarlet, corolla dark purple. Flowers small but very free. Growth is dwarf and spreading. Has been used for path edging as its habit of growth helps to disguise the sharpness of the path.

'El Cid' (Colville 1966). Single. Tube and sepals bright red. Corolla rich burgundy. Well-formed, medium-sized flower produced in profusion. Strong upright growth. 'Old fashioned' in appearance but none the less charming for that.

'Eleanor Rawlins' (Wood 1954). Single. The tube and the long reflexing sepals are a bright spinel red. The corolla is dark lake. Flowers are of medium size and very free. Growth is sturdy, upright and bushy. An extremely good, modern hardy variety.

'Elfin Glade' (Colville 1964). Single. The tube and sepals are a salmon-rose and the corolla is of a deeper shade of the same colour. The small flowers are borne in profusion and are similar to that well-tried hardy 'Margaret Brown', except being slightly deeper in colour and a little larger. Quite hardy and will make a medium-size bush outdoors.

'Empress of Prussia' (Hoppe 1868). Single. Tube and sepals bright glowing scarlet, corolla scarlet lake. For a hardy variety the flowers are quite large and most prolific. Growth is short but sturdy and upright. This variety was lost for years and rediscovered in 1950 growing outdoors in a west country garden, where it had been established for fifty years. One of the best of the recognized hardy varieties.

'Florence Turner' (Turner 1955). Single. Tube is china rose which pales down to white on the top of the sepals. The underside of the sepals is white shaded with magnolia-purple. Corolla is magnolia-purple. Growth is upright and bushy. Given good cultivation, and a sheltered spot, this variety will make a very handsome small bush. Very free flowering and very hardy.

'Frau Hilde Rademacher' (Rademacher 1901). Double. The tube and sepals are red. The petals of the corolla are blue with cerise veining and they turn to a rich lilac as the flower matures. Growth is bushy and upright and the small flowers are borne in some profusion. An excellent hardy cultivar for which enthusiasts are thankful for its reintroduction to general cultivation.

'General Voyron' (Lemoine 1901). Single. Tube and sepals cerise, corolla violet magenta. Flowers are of medium size and produced in good numbers. Growth is upright with long arching fronds. Has been proved hardy in many parts of Great Britain. A very handsome fuchsia and a great favourite of the author. Under greenhouse conditions will make a fine standard.

'Graf Witte' (Lemoine 1899). Single. Tube and sepals scarlet, corolla purple but shaded with a rosy mauve. Flowers are very small but very freely produced. Growth is upright, of medium size and bushy. Foliage is a yellowish green. It would appear to be hardy in most parts of the country.

'Herald' (1869). Single. The tube and the gracefully recurving sepals are scarlet-cerise, the corolla is dark purple. Flowers are of medium size and very free. Growth is upright and bushy. This variety is inevitably cut down by winter frosts, but soon recovers in the spring to make a neat little bush.

'H. G. Brown' (Wood 1949). Single. Tube and sepals deep scarlet, corolla Indian lake. Flowers are small to medium and produced in profusion. The foliage is dark green and has a distinctly holly-like appearance. Growth is low and bushy. Very hardy.

'Holstein' (Schadendorff 1912). Single. The tube and the long recurving sepals are cerise, the corolla has petals of a dull mauve suffused with pink. Flowers are large for the hardy section, and very free. Growth is rather straggling in habit and very bushy. Given a nice sheltered spot in the garden, it will make a sprawling mat of colour.

'Howlett's Hardy' (Howlett 1952). Single. Tube and sepals scarlet, corolla purple. The flowers are large for a fuchsia of such hardiness. Growth is medium and bushy. Very free flowering. One of the last introductions of a great plantsman who kept the name 'fuchsia' alive during the period between the wars.

'Joan Cooper' (Wood 1954). Single. Tube and sepals rose opal with a definite green tip to each sepal. The corolla is described as cherry-red. The flower is quite distinctive as the sepals reflex straight back and cover the tube. Although the flowers are rather small they are produced in profusion. Growth is upright and rather open. An unusual-coloured flower for a hardy variety.

'Joan Smith' (Thorne 1959). Single. Tube white, sepals flesh pink. Corolla soft pink. The medium-sized flower is produced in profusion. The sepals reflex flat to the tube, completely covering it. Growth is strong, upright and very fast growing. Ideal for bush, half standard and standard growing in the greenhouse.

'John' (Wood 1949). Single. Tube and sepals dull crimson, corolla mauve-cerise. The flowers are very small but not so freely produced as some in this class. Growth is dwarf but erect. Very hardy and at home in the rockery.

'Kathleen' (Wood 1949). Single. Tube and sepals cherry-red, the sepals tipped with greenish yellow, corolla rose-pink. Flower very small but quite free. Growth is erect and dwarf. Very hardy.

'Lena' (Bunney 1862). Semi-double. Tube and sepals flesh pink, corolla orchid purple. The sepals half reflex to stand away from a rather spreading corolla. Flowers are of medium size and extremely free. Growth is lax but capable of being trained to any required shape. This is undoubtedly one of the most versatile varieties now grown, for besides being quite hardy, in greenhouse culture it will make a fine basket, standard, espalier and, with patience, a pyramid.

'Lottie Hobby'. Single. Tube is dull scarlet, sepals are tipped rose. The petals of the corolla are scarlet and flatten themselves out. The flowers can only be described as tiny, but they grow in good numbers. The holly-like foliage is small, dark green and glossy. Growth is short but upright. This is one of the few 'Breviflorae' hybrids of the 'Encliandra' section. Dainty, attractive and extremely hardy.

'Margery Blake' (Wood 1950). Single. Tube and sepals scarlet, corolla solferino purple. Flowers are very small but borne in profusion. Growth is that of a low-spreading bush. Has a very long season of flower. Very charming.

'Mrs. W. P. Wood' (Wood 1949). Single. Tube and sepals pale pink, corolla pure white. Flowers are very small but most profuse. Growth is stiff and upright. In milder parts of the country this variety will make a fine shrub, but its hardiness is suspect in colder districts. A hybrid from *F. magellanica molinae* which it resembles, except that the flowers are larger and it is definitely not so hardy as its parent.

'Peggy King' (Wood 1954). Single. Tube and sepals claret rose, the slightly open corolla is peony purple. Flowers are small and very freely produced. Growth is sturdy, upright and very bushy. A good hardy variety.

'Pixie' (Russell 1959). Single. Tube and sepals scarlet, corolla mauve. Flowers are very small but profuse. Growth is upright, of medium size and bushy. Foliage is a yellowish green. This variety is a sport from the well-known hardy variety 'Graf Witte'. A good colour break.

'Prince of Orange' (Banks 1872). Single. Tube and sepals salmon-pink, corolla orange-vermilion. The sepals hang over the corolla. Flowers are medium to large and very free. Growth is upright and bushy. The colouring is most distinctive and gives the impression that this variety is very delicate. It has, however, a much hardier constitution than it has been previously credited with.

'Prostrata' (Scholfield 1841). Single. Tube and sepals dull red, corolla violet. Flowers are very small but borne in profusion. Growth as the name implies is low and spreading. Ideal variety for the rockery where it will make a neat mat of colour.

'Pumila' (Young 1821). Single. Tube and sepals scarlet, corolla mauve. Flowers are small but profuse. Growth is that of a miniature bush. Makes a charming rockery subject and is often displayed as such on the delightful rockery displays at Chelsea Flower Show and R.H.S. Shows.

'Purple Cornelissen' (Cornelissen 1897). Single. Tube and sepals cerise, corolla dull purple. Extremely free flowering, the blooms being of medium size. Growth is lower and more dense than its famous relation 'Madame Cornelissen'. Quite hardy, but flowers too late in exposed areas. Given a place in the shelter of the shrubbery, this variety will flower well.

'Rhombifolia' (Lemoine). Single. Tube and sepals scarlet-cerise, the sepals overhanging the short purple corolla. Flowers are quite small but are very profuse. Growth is upright, strong and free branching. Recorded as a seedling from F. Riccartonii but it is quite distinct from its parent.

'Ruth' (Wood 1949). Single. Tube and sepals rosy red, corolla purplish red. Flowers are of medium size and bloom profusely. Growth is upright and bushy. Very hardy and quite attractive.

'Schneeball' (Twrdy 1874). Double. The tube and sepals are rose-red. The corolla is white, heavily veined pink. For a hardy cultivar the flower is quite large. Very free flowering on a bushy upright shrub. This cultivar has been kept in existence by some enthusiast and only recently found its way back into commerce.

'Schneewittchen' (Hoech 1884). Single. The tube and sepals are red. The corolla is rich violet, heavily veined red. A very vigorous, self-branching and free-flowering cultivar which has proved itself hardy in the south of England. It could well be hardy in the north. The flower is large for a hardy. Gives a wonderful show in the garden. Grown as a standard for summer bedding it is ideal.

'Silverdale' (Travis 1962). Single. A *F. magellanica* seedling with ivory tube and sepals, flushed pink, and pale lavender corolla. Very similar in flower and habit to *mag. alba (F. m. molinae)*, but flowers nearly twice the size. A useful addition to the hardies.

'Star Light' (Bull 1868). Single. The waxy textured tube and sepals are creamy white, the petals of the corolla are rosy cerise with white at the base. Flowers are of medium size and borne in profusion. Growth is upright and bushy. Although not sold as a hardy variety it has a tough constitution, and many plants survived outdoors during the severe winter 1962–3. An 'easy to grow' plant and worth experimenting with.

'Susan Travis' (Travis 1958). Single. Tube and sepals deep pink, corolla rose-pink. For a variety so hardy the flowers are quite large and very freely produced. Growth is upright, medium and bushy. Makes a grand bedding fuchsia that can be left outside permanently.

'The Tarns' (Travis 1962). Single. The tube is pale pink and the sepals are pale pink on the upper surface but a deeper pink beneath. They have a green tip. The corolla is violet-blue, paling to rose, and then to white at the base of the petals. A very free-flowering cultivar which when grown as a hardy fuchsia makes a low, spreading, and free-branching bush. Has a larger flower than most hardy cultivars. It has proved its hardiness in the south of England and some places in the north. Will also make a fine standard in the greenhouse.

'Three Counties' (Raiser unknown). Single. The tube, and the sepals, which turn right back, are scarlet-cerise, the corolla is bluish violet. The flowers are medium to large and very free. Growth is upright and bushy. Given protection from cold winds this variety will flourish in the open. Named after the show where it was seen by a leading fuchsia nurseryman.

'Tom Thumb' (Baudinat 1850). Single. Tube and sepals light carmine, corolla mauve-purple. A medium-sized flower borne in great profusion. Growth is dwarf and upright. Good rock and dwarf-hedging plant. Very popular.

'Trase' (Dawson 1956). Double. Tube and sepals cerise, corolla white with carmine veining on the petals. A most prolific bloomer with flowers on the medium to small size. Growth is strong, upright and bushy. Very hardy and a delight to grow. Becoming extremely popular.

'Try Me O' (Banks 1868). Single. The tube and the reflexed sepals are carmine, the open corolla a rosy magenta. Flowers of medium size are freely produced. Growth is dwarf and spreading. Although not recognized as a true hardy variety it has been proved that if sited well, it will last indefinitely in the open ground.

'W. P. Wood' (Wood 1954). Single. Tube and sepals rich blood-red, corolla royal purple with the petals tipped violet-purple. The blooms are of medium size and very free. Growth is dwarf, upright and very bushy. Makes a fine bedder.

CHAPTER XVII

Varieties Suitable for Growing
as Standards

'Amelie Aubin' (Eggbrecht 1884). Single. The tube is of waxy appearance, long and thick. The sepals are also waxy and are upswept. Both are a creamy white in colour. Corolla rosy carmine. The large flowers are very free. Growth is strong, upright and bushy. Makes a grand standard. Some think the flower lacks the grace of the majority of fuchsias, but it is an excellent variety for the display it will give. Can also be grown well in bush form and pyramid form.

'Amy Lye' (Lye 1885). Single. The stout tube and the broad sepals are white, of waxy appearance, with the very faintest pink tinge in them. The corolla is rosy cerise. Growth is vigorous and upright. The oval leaves are of a medium green. The flowers are of medium size and most profuse. Makes an excellent standard but it should be grown with a fairly long head to get the best effect. Can, with good cultivation, be grown to pyramid shape.

'Beauty of Trowbridge' (Lye 1881). Single. The waxy-looking tube and recurved sepals are a creamy white. The corolla rosy cerise. A free-blooming variety having flowers of medium size. Growth is strong and erect. Should be grown as a standard with a short head, as it is a variety that will take a lot of pinching back. A very popular standard variety. Also good in bush form.

'Elizabeth' (Whiteman). Single. The long tube and sepals are rose-pink, the corolla salmon-pink. The blooms are large and extremely free. Growth strong, vigorous and upright. Grow with a short head, for the vigorous growth will need much pinching to prevent the head getting unwieldy, and to get the maximum flower. A distinctive variety always admired. Has also excelled as a greenhouse climber.

'Falling Stars' (Reiter 1941). Single. Tube and sepals scarlet, corolla turkey red. A prolific bloomer having flowers of a medium size. Growth is vigorous and of a spraying habit. Best effect is gained by growing with a long head. Besides making a good standard this variety is ideal for bush growing and also for basket work. An outstanding American introduction.

'Gustave Doré' (Lemoine 1880). Double. Tube and sepals pale pink, corolla pure white. Very free flowering with well-shaped medium-sized flowers. Growth is vigorous and upright. Very easy to grow and can be had in flower quite early in the season. An old variety which is well able to hold its own against the modern introductions. Very gay. Will also make a fine bush plant.

'Joy Patmore' (Turner 1961). Single. Pure waxy-white tube and sepals. Corolla flared, a rich carmine with white base to each petal. The flowers are of medium size and profusely borne. Growth is strong and upright. A good variety for either bush or standard.

'La France' (Lemoine 1885). Double. Tube and sepals scarlet, the very full corolla violet-purple. Flowers are large and free. Growth is strong, upright and bushy. Best grown with a longish head as too much pinching will, in this case, encourage the growth of small shoots at the expense of flower. Ideal also in bush form.

'Mrs. Lovell Swisher' (Evans and Reeves 1942). Single. The tube is flesh pink shading to the sepals which are ivory white. Corolla pink to deep rose. Flowers are small but borne in great profusion. Growth is vigorous and upright, but it is not a vigour that will get out of control. Has an extremely long period of flower. Suitable also for growing in pyramid and bush forms.

'Mrs. Rundle' (Rundle 1883). Single. The long tube and slender sepals are flesh pink, corolla orange-vermilion. The flowers are long and very graceful. Growth is vigorous and inclined to be lax. Prolific in bloom. This variety will make the finest weeping standard. Constant attention to tying is needed, right from the cutting stage, to get a straight stem. Grow to a short head as the head growth will arch out gracefully and then almost cascade. Suitable for hanging baskets or for growing in bush form.

'Pepi' (Tiret 1963). Double. The tube and the broad incurving sepals are rosy red. The large, full, double corolla is orange-red which matures to a light bronze. An attractive flower freely produced on an upright growing plant. As a standard it will make a full head as the upright branches are weighed down by the weight of the blooms. Easy to get that nice straight stem to support the head. It will also make a good bush.

'Pink Temptation' (Wills 1966). Single. The tube and sepals are creamy white and the corolla is of pale Tyrian-rose. A very free-flowering sport of that popular cultivar 'Temptation', to which it is identical except for the colouring. Easy to grow as a standard and the natural bushy habit of this cultivar means that no trouble should be experienced in forming a neat, thick, floriferous head.

'Queen Mary' (Howlett 1911). Single. The tube is pink and the sepals are of the same colour but tipped white. Corolla rose-pink passing to a deep mauve. Flowers are large and very free. Growth is sturdy and upright. A variety that will stand a lot of rough handling. Equally suitable for growing in bush form.

'Rose of Castile Improved' (Lane 1871). Single. Tube and sepals flesh-coloured, corolla violet-purple. Flowers are large and very free. Growth is vigorous and upright. It will make a very large standard, a large bush, or a first-class climber in the greenhouse. A gross feeder requiring plenty of moisture.

'San Mateo' (Niederholzer 1946). Double. Tube and sepals are pink, corolla dark violet with splashes of pink on the petals. The blooms are very large and very freely produced. To make a fine standard frequent attention should be made to tying as growth can be lax. As a standard this variety is capable of making a large head, as growth is fairly upright until the flowers appear, when the extra weight will make it suddenly cascade. Grow also as bush or as a trailer in hanging baskets.

'Santa Cruz' (Tiret 1947). Double. Tube and sepals crimson, the corolla being of the same colour, but a darker shade. Flowers are large and very freely produced. Growth is extremely strong and upright. The large foliage has distinct red veining. A variety which must be grown with a long head, as growth is stiff and there will be no arching of growth to give the appearance of depth. Plenty of pinching out will give a head that is upright and bushy. A variety which has substance both in growth and flower. Will also make a very large bush.

'Streamliner' (Tiret 1951). Single to semi-double. The long thin tube, and the very long, twisted and curved sepals are crimson. The corolla is also long and is a deeper shade of crimson. The whole flower is essentially long rather than bulky. Very free bloomer. Growth is very lax and constant attention to tying is necessary. Grown with a short head, with several pinches, this variety makes a fine weeping standard. A good variety also for hanging baskets. Suitable also for espalier work.

'Whitemost' (Niederholzer 1942). Single to semi-double. Tube and sepals are of an extremely pale pink, corolla white with the merest touch of pink in it. Flowers are quite large and very free, and the ruffled petals make them look more double than they really are. Growth is strong and upright. A variety that can be grown with a medium head as it is naturally quite bushy and will form a neat bush at the head. Excellent also in bush form.

'Whittier' (Home Gardens 1955). Double. Tube and sepals white, the tips of the sepals being tipped green. The corolla is white, very faintly flushed pink, with a surrounding crown of very pale pink petaloids. The flowers are large, and, for their size, most profuse. Grow with a fairly short head as the weight of bloom will bring a graceful arch to the branches. Ideal also for growing in bush form.

CHAPTER XVIII

Varieties Suitable for Espalier

'**Chang**' (Hazard 1946). Single. The tube and the neatly recurving sepals are orange with red tips to the sepals. The corolla is a most brilliant orange. The flowers are small but most prolific. Growth is strong, upright and free branching. Very aptly named for this variety has a definite oriental appearance to it. Not a variety that everybody can grow successfully, but one that all should try. Makes, also, a charming bush or half standard.

'**Coquette**' (Niederholzer 1945). Single. The tube is white, sepals red, and the corolla purple. Blooms are of medium size and extremely free. Growth is strong, spreading and free branching. An easy plant to grow and ideal for espalier work. Good also in bush form.

'**Hebe**' (Stokes 1848). Single. Tube and sepals are pure white, the sepals are held at a good angle. The corolla opens violet and ages to a rich reddish crimson. The medium-sized flowers are borne in great profusion. Growth is willowy and the medium green foliage has a deeply serrated edge. Certainly one of the most free flowering of fuchsias. Eminently suitable for espaliers.

'**Miss California**' (Hodges 1950). Single to semi-double. The tube and the upturned sepals are a pale pink, the corolla is long and is white with a faint pink flushing inside the petals, and pale pink veining. The medium to large flowers are very freely produced. Growth is strong but lax, and easily trained. The delicate colourings are always admired. Will also cascade for basket work.

'Morning Mist' (Berkeley Horticultural Nursery 1951). Single. The long tube and sepals are orange-rose, the corolla orange-red, suffused purple. A prolific bloomer with distinct long flowers. Growth is strong but lax, making it an easy variety to grow as espalier. With patient training will make a fine weeping standard.

'Mrs. J. D. Fredericks' (Evans and Reeves 1936). Single. Tube and sepals salmon-pink, corolla a deeper shade of pink. The small flowers are borne in great clusters on a vigorous upright bush. The foliage is a light green, and acts as a perfect foil to the numerous flowers. Grow also as standard or bush.

'Pee Wee Rose' (Niederholzer 1939). Semi-double. Tube and sepals red, corolla clear rose. The flowers are small but are borne in great numbers. The foliage is very small. It is a natural trailer but is easy to train providing sufficient ties are given. It is a vigorous grower and in the first year will make a grand plant. Ideal also for hanging baskets.

'Powder Puff' (Hodges 1953). Double. Tube and sepals Tyrian rose, shading to rose madder at the tip of heavy recurved sepals. Corolla clear pink. Flowers are medium to large, and very free. A vigorous, lax-growing variety which by nature of its habit lends itself to espalier training. An extremely attractive variety. Suitable also for hanging baskets.

'Rubeo' (Tiret 1947). Semi-double. The waxy-looking tube and sepals are carmine, corolla rose bengal at the centre, shading to a rich crimson with flecks of dark lake. The large blooms are very free. Growth is upright and bushy. Best trained as a fan-shaped espalier. Also makes a good bush.

'San Francisco' (Reiter 1941). Single. Tube and sepals rosy carmine, corolla bright geranium lake. The tube is very long making a long-looking flower. Very free blooming. Growth is vigorous but lax. Easy to train as an espalier. Will also make a fine weeping standard. The foliage is thick and large.

'Sierra Blue' (Waltz 1957). Double. The short tube is white, the sepals are white flushed inside with the palest pink. Corolla

235

is very full and in colour is of silver-blue, changing with age to a soft lilac. Flowers are quite large and freely produced. Growth is upright with long arching branches. A very lovely flower. Will make a grand bush plant.

'Swing Time' (Tiret 1950). Double. The tube and short, up-turned, crêpe-textured sepals are a rich red, the corolla is a milky white with slight pink veining at the base of the petals. The flowers are large and extremely free. Growth is upright and free branching. A fine-shaped flower. Can be grown also in standard or bush form.

'Thunderbird' (Tiret 1957). Double. Tube and sepals rose, corolla vermilion to china rose. The flowers are large and freely produced. Growth is cascade and self-branching. With careful training this variety makes a wonderful espalier. Worth trying. Looks wonderful also in a hanging basket.

'White Spider' (W. Haag 1951). Single. Tube and long twist-ing sepals are a baby pink, the single corolla is a clear white. The flowers are long and large and most prolific. A strong fast grower, this variety, besides making a fine espalier, is equally good as a bush or weeping standard, and is an excellent basket plant.

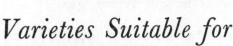

CHAPTER XIX

Varieties Suitable for Hanging Baskets

'Ailsa Garnett' (Thomas 1965). Double. The tube and sepals are white. The corolla is garnet-red. The large flowers are freely produced on a natural basket-type plant. It is self-branching and quite vigorous in its habit of growth. It is a lovely flower and can best be appreciated by being viewed from below, as it would be hanging in a basket.

'Balkon' (Neubronner). Single. Tube and sepals pale pink, corolla deep pink. Flowers are of medium size and borne freely. Growth medium cascade. Ideal for small baskets and covering the front of window boxes and the greenhouse staging. One of the few basket varieties remaining from the pre-1914 days, the rest having given way to the modern varieties from America.

'Berkeley' (Reiter 1955). Double. Tube and sepals pale rose, corolla very dark rose-red, edged with pale rose. The flowers are large and quite prolific, being borne in large clusters. Growth is cascading and free branching. In the large 16-inch basket five plants are usually necessary to get the best effect. Recommended.

'Bon Bon' (Kennett 1963). Double. The tube is white and long and the short broad sepals are pink. The tight corolla is a very pale pink. The flowers are of medium size and fairly free, and look well against the small glossy foliage. Trailing in habit but should make a good weeping standard.

'Bouffant' (Tiret 1948). Single. Tube and sepals red, corolla white heavily flushed with rose. The flowers are long with the corolla opening wide. Very prolific bloomer with quite large flowers. Will make a nice medium-sized basket.

'Bridal Veil' (Waltz 1963). Double. The flowers are a creamy-white self and are produced on a plant with lax growth and small glossy foliage. It is self-branching and most floriferous. One of the best of the white basket types.

'Bubble Hanger' (Niederholzer 1946). Single. The tube and the spreading sepals are pale apricot-pink. The large petals of the corolla are a soft transparent rose. The flowers are large and extremely free. Growth is fast, vigorous and naturally cascading. The name is derived from the conspicuous, fat buds, which really do look like bubbles. Will make a nice basket very quickly.

'Butterfly' (Reiter 1942). Single. The tube and the long recurved, twisting sepals are crimson, the corolla rose bengal. The large petals of the corolla also curve back and twist slightly. The buds are long, slender and pointed. Flowers are large and very free and show up well against the fine sawtooth-edged foliage, which is a light green in colour. Makes a most attractive basket.

'Caballero' (Kennett 1965). Double. Tube and sepals deep salmon-pink. Corolla is bluish purple to violet with distinct splashes of salmon-pink. The petaloids which surround the corolla vary from white to red. A large striking flower and freely produced. Will make a striking basket.

'Cara Mia' (Schnabel 1957). Semi-double. Tube and sepals are pale rose, the corolla is dark crimson. The flowers are very large and very prolific. In the larger baskets five plants might be better than the usual four. A basket of this variety in full bloom is a beautiful sight.

'Cascade' (Lagen 1937). Single. The tube and the long, slender sepals are whitish, heavily flushed with carmine; the corolla is deep carmine. The flowers are medium to large and borne in great profusion. A vigorous grower which will make a large basket quickly. An extremely popular variety which has almost become the 'yardstick', against which the merit of other basket types is judged.

'Cathie MacDougall' (Thorne 1960). Double. The tube and sepals are a rich pink and the full corolla has petals which are mottled mauve and pink. The flower is large and is produced in quantity. It is a vigorous grower and in a 16-inch basket four plants should be ample to make a full and attractive basket. With a little training it can make an attractive bush plant.

'City of Millbrae' (Martin 1958). Double. Tube and sepals bright pink with the sepals being just tipped green. The very full corolla is orchid blue. The blooms are quite large and free. Growth is natural cascade.

'Claret Cup' (Lagen 1940). Single. Tube and sepals very pale pink, corolla rich claret. This is the sister seedling of the famous basket variety 'Cascade'. In many ways it is similar to that variety, the main difference being the slightly larger flower of 'Claret Cup' and also the more intense colourings.

'Crackerjack' (Fuchsia-La Nursery 1961). Single. Tube and sepals white, faintly flushed pink. Corolla pale blue ageing to pale mauve. Flower is large with very long petals. Free flowering and will make an attractive basket.

'Creole' (Schnabel 1949). Double. Tube and sepals are crimson, the corolla, consisting of four long centre petals surrounded by many smaller petals, is a dark crimson splashed with maroon and crimson. Flowers are large and free. A very flashy variety that will always attract attention.

'Ebb Tide' (Erickson-Lewis 1959). Double. The tube and recurved sepals are white outside and pale pink inside. The spreading corolla is of light blue and phlox-pink, and the light blue changes with age to a pastel lavender giving the effect of two different varieties growing from the same shrub. Flowers large and freely produced on a vigorous-growing plant. An excellent basket type.

'Golden Marinka' (Weber 1955). Semi-double. The tube and sepals rich red, corolla slightly darker red. Almost a self colour. Flowers are of medium size and most prolific. Quite vigorous and can be trained to most other shapes, although when grown naturally it will cascade. This variety is a sport of the popular variety 'Marinka', the main difference being the foliage which is light green, barred and splashed creamy gold. Seems to lack the sustained vigour of its parent.

'Interlude' (Kennett 1960). Double. The tube is slender, and the sepals are waxy white with pink flush. Corolla has four basic petals, violet-purple, with secondary outer petals a delicate orchid-pink. Stamens and stigma are beautifully prominent on mature flowers. Flowers very free and of medium size. A basket variety.

'La Campanella' (Blackwell 1968). Double/Semi-double. The tube and sepals are white. The corolla opens as imperial purple and changes to lavender as the flower matures. The flowers are most prolific and are borne on a plant which is self-branching and of trailing growth. This cultivar makes a basket of outstanding beauty and is undoubtedly one of the most outstanding basket plants of recent introduction. A rapid grower.

'Lilibet' (Hodges 1954). Double. The long tube and recurving sepals are white flushed pale carmine, and the corolla is soft rose flushed with geranium lake. Flowers are large and very free. The buds are long and pointed and are themselves attractive. An ideal basket variety.

'Lullaby' (Reiter 1953). Double. The tube and recurved sepals are white, faintly tinged with pink, the corolla is a rich lavender-pink. Flowers are medium to large and very free. Growth is natural and cascade. The pastel colourings of this variety make it extremely attractive.

'Mantilla' (Reiter 1948). Single. Tube, sepals and corolla self-coloured, deep carmine. The shape of the flower is quite distinct, consisting of a long narrow tube, very small sepals and a small corolla. These are borne in clusters. The foliage has a bronze hue to it. Needs that extra plant to make a full basket but it is extremely unusual and worth growing.

'Niobe' (Reiter 1950). Double. The tube and upturned twisting sepals are white flushed pale pink, the corolla is an intense smoky rose colour. The anthers often stand outside the corolla. An extremely fast-growing cascade which will make a superb basket very quickly. A leading variety for basket work.

'Omeomy' (Kennett 1963). Double. The long sepals and tube are pale pink. Large, double, tight corolla of dianthus purple is overlaid with coral-pink marbling. The large flowers are very freely produced and stand out well against the medium green foliage. A natural trailer, it is ideal for hanging baskets.

'Pastel' (Reiter 1941). Single. Tube and sepals a delicate pale rose, corolla pale lavender-blue. The sepals reflex in a unique curl. Growth is cascade but not very vigorous. Definitely not suitable for exhibition as the plants in a single basket are usually restricted to four in number and this variety needs five or six plants in a 16-inch basket. The flower is however one of the most beautiful among the fuchsia hybrids.

'Pebble Beach' (Niederholzer 1947). Single. Tube is white, the upturned sepals are pale rose, and the corolla is dark purple. Flowers are large and freely produced. Growth is naturally of a cascade habit. Very attractive.

'Red Jacket' (Waltz 1958). Double. Tube and sepals bright red, corolla pure white. The corolla is large and fluffy and the very long stamens and pistil are dark red. Flowers are large and profuse. Growth is vigorous cascade. A fairly recent introduction that cannot fail to become popular when it becomes better known.

'Red Spider' (Reiter 1946). Single. The long tube and narrow recurving sepals are deep crimson, corolla deep rose madder, veined and margined deep crimson. Flowers are long, large and very prolific. Growth is vigorous cascade, and it will quickly make a good basket.

Varieties Suitable for Hanging Baskets

'Trail Blazer' (Reiter 1951). Double. The long tube and the recurved sepals are magenta, the corolla is a darker shade of magenta. The flower is large and long and most profuse. Growth is extremely vigorous in a natural cascade habit. This variety will quickly make a fine basket. Very popular and quite worthy of that popularity.

'Trailing Queen' (European). Single. Tube and sepals light red, corolla magenta. Flowers are medium to small and very free. Growth is cascade and very vigorous. The pale green foliage shows up the flowers to advantage. Makes a good basket.

CHAPTER XX

Varieties Suitable for Growing as Pyramid

'**Achievement**' (Melville 1886). Single. The tube and the gracefully recurved sepals are cerise, the corolla is purple. Flowers are medium to large, and most profuse. The medium-green foliage is slightly serrated. A rapid, upright growing variety, which is suitable for training to any shape. Can be had in bloom very early in the season. An old variety which has never been surpassed in its colour class.

'**Brutus**' (Lemoine 1897). Single. The tube, which is short, is a rich cerise, as also are the sepals, which recurve gracefully. The corolla is a rich purple. It has a most vigorous habit of growth, and it covers itself with medium-sized flowers. The foliage is a pronounced green with dark veins. Besides making a grand pyramid it will excel grown as a standard or bush.

'**Checkerboard**' (Walker and Jones 1948). Single. The long tube is red, the slightly recurved sepals are red where they meet the tube, and change abruptly to white. The corolla is red, but deeper and brighter than the tube. Very heavy blooming with flowers that are long and of medium size. Growth is vigorous, upright and bushy. A well-flowered pyramid of this variety is one of the most spectacular plants of the fuchsia world. Naturally showy, the myriads of flowers produced in such mass would stand out for a considerable distance. Very aptly named. Grow also in standard or bush form.

'**Constance**' (Berkeley Horticultural Nursery 1935). Double. Tube and sepals pink, corolla a medium shade of mauve with

243

faint pink markings. Medium-sized blooms very freely produced. Growth is strong and upright, and easy to train to any required shape. An American sport of the variety 'Pink Pearl'. A fine variety for the beginner who wishes to grow pyramids.

'Marinka' (Rozain-Boucharlat 1902). Single. Tube and sepals are rich red, and the corolla is also a rich red, but a little darker in shade. Almost a self-coloured flower. A most profuse bloomer with flowers of a medium size. Growth is naturally lax, and when training as a pyramid all growth must be trained upright until the shape is attained, when it can be allowed to cascade. There is no shape to which this versatile variety cannot be trained successfully. Very popular for hanging baskets, standards, espaliers and bush, as well as pyramid.

'Molesworth' (Lemoine 1903). Double. Tube and sepals scarlet-cerise, the full corolla is white with light pink veining at the base of the petals. The shapely flowers are of medium size and very freely produced. Growth is strong, upright and very bushy. Foliage is a sage-green in colour. The growth can become too dense with this variety and it may be necessary to remove secondary shoots here and there, to allow air to circulate through the pyramid. Will make a fine, thick pyramid. Grand for growing in bush form or as a standard.

'Mrs. Marshall' (British). Single. The waxy-looking tube and sepals are creamy white, the corolla is rosy cerise. Flowers are of medium size and are borne profusely. Growth is strong and upright. Although not so vigorous as others mentioned in this section it will make a good pyramid of fair proportions. Although the origin of this fuchsia is not known, it has been sold in the market places and general stores of Great Britain since the beginning of the century. A variety suitable for training to all known shapes.

'Pink Pearl' (Bright 1919). Semi-double to double. Tube and sepals pale pink, corolla a deeper shade of pink. The blooms are of medium size and very freely produced. Growth is vigorous and upright. This variety was raised by one of the greatest exponents of the art of growing pyramids. A lovely fuchsia and one that is easy to grow. Like its sport 'Constance' this is one that should be tried by the beginner. Good also in bush or standard form.

'Royal Purple' (Lemoine 1896). Single. The waxy tube and sepals are cerise, the corolla a most vivid royal purple. The flowers are large and very free. Growth is strong and upright. The intensity of the colours will always make this fuchsia stand out. Besides being ideal for pyramid work, this variety made its name as a climber in the No. 4 Temperate House at Kew where, before that greenhouse was rebuilt, it was admired by thousands of visitors. Good also for bush or standard.

'Snowcap' (Henderson). Double. Tube and sepals red, corolla pure white. A most profuse bloomer having medium-sized flowers. Growth is vigorous and upright. Undoubtedly one of the freest-flowering double varieties. Both the habit of growth and the quantity of flower make this an ideal fuchsia for pyramid work. Ideal also for growing as bush or standard.

CHAPTER XXI

Varieties Recommended for Exhibition

'Alaska' (Schnabel 1963). Double. Tube white, sepals white with faint pink tinge, corolla white and puffy in shape. The flowers are large and fairly free, and show up well against the medium green foliage. Easy and attractive.

'Alfred Rambaud' (Lemoine 1896). Double. Tube and sepals scarlet, corolla very rich purple, ageing to rosy purple. Flowers are large and very free, and the corolla open and very double. Growth is upright and bushy. Has one fault, which is that the colour of the flowers gives away their age at a glance. If too many flowers have the rosy purple corolla, this may fail to please the judges. Grow as bush or standard.

'Alison Ryle' (Ryle 1967). Double. A beautiful seedling from the ever popular 'Tennessee Waltz', which seems to have practically all the excellent attributes of its famous parent. The main difference would be that the corolla is a deeper tone of lilac-lavender and the petals are more flared. Suitable for all forms of training.

'Angela Leslie' (Tiret 1960). Double. Tube and sepals are pale flesh-pink, the very full corolla bright pink. The flower is enormous, and for such a size is quite free. Growth is strong, upright and bushy. An extremely showy variety. Full of glamour and sure to attract attention.

'Artus' (Crousse). Double. Tube and sepals rosy scarlet, corolla pure white. The corolla has a very square appearance which

246

makes it unique in its colour section. Flowers are large and free. Growth is upright and bushy. The unusual shape of the flower always attracts attention.

'Bella Forbes' (Forbes 1890). Double. Tube and sepals cerise, corolla creamy white. The flower can be described as 'classic' in shape, the sepals standing out to reveal the tight double corolla. They are of medium size and very freely produced. Growth is vigorous and upright. Easy to grow, but given the extra attention exhibiting demands, this can be a superb bush. Also makes a nice standard.

'Bewitched' (Tiret 1951). Double. Tube and sepals white, the sepals being tipped green. The corolla is deep purple shading to white at base, the deep purple changing to violet as the flower develops. The flower is large and very free. Growth is vigorous, but lacks the substance to support the wealth of heavy blooms. With careful and discreet tying this variety can be a winner. Left to trail it will make a fair hanging basket.

'Bridesmaid' (Tiret 1952). Double. The tube and sepals are white with a pale carmine blush to them. Corolla pale lilac pink, darkening at edge of petals. The flower is large and very freely produced. Growth is upright and bushy. The bloom is neat and showy and well set off by the tidy habit of the growth.

'Bunker Boy' (Tiret 1952). Double. Thick waxy white tube, sepals strong, upturned, white flushed pale carmine. The long corolla has fluted petals which are white at the base, deepening to geranium-lake at the edges, surrounded by a circle of petaloids which are pink. Flowers are large and free. Growth is strong and upright. Very attractive variety which when well grown cannot fail to attract. Should also be tried in standard form.

'Carmel Blue' (Hodges 1956). Single. The long narrow tube is greenish white and the outspread sepals are white flushed with the palest blush on the underside and tipped green. The long corolla is a smoky blue. Growth is vigorous and upright and carries a multitude of flowers which show up well

247

against the excellent foliage. Ideal for standards, espaliers, and pyramids. A worthy fuchsia.

'Citation' (Hodges 1953). Single. Tube and sepals rose-pink, corolla white with pink veining at the base of the petals. The sepals curl right back and cover the short tube. The corolla consists of four broad petals which open out campanulate. A prolific bloomer with large flowers. Growth is upright and well branched. Constantly admired, this is undoubtedly one of the most popular fuchsias of today. Ideal for growing as bush or standard. Easy to grow and always in flower.

'Claire Evans' (Evans and Reeves 1952). Tube white, sepals white on upper surface, shell-pink underneath, corolla light violet-blue fading to rosy mauve with age. Flowers are large and free and very crisp-looking. Growth is medium, upright and very bushy. Very easy to grow and train and is ideal for growing for the earlier shows, as this variety can be brought on to flower quite early in the season.

'Collingwood' (Niederholzer 1945). Double. Tube and sepals pale pink, corolla almost pure white, with the faintest tinge of pink in the petals. The flower is very large, and for its size very freely produced. Growth is strong, upright and self branching. This is an easy variety to grow for the show table for it requires little training. Well grown this variety cannot fail to catch the judges' eye. Will also make a fine standard.

'Connie' (Dawson 1961). The tube and sepals are turkey-red, the corolla off-white veined and flushed pink. The medium-sized flowers are produced in some quantity. Growth is strong, upright, and naturally bushy. Remains in flower for a long period. Has already won the Jones Cup for being the outstanding seedling of its year.

'Court Jester' (Tiret 1960). Double. The short tube and the thick crêpe sepals are deep rose, the large corolla is royal purple, with pink-coral petals that overlap, giving the effect of a jester's costume. Growth is strong and upright, and the flowers large and plentiful. A novelty sure to attract attention.

'Crescendo' (Reiter 1942). Semi-double. Tube and sepals turkey-red, corolla turkey-red with purple flushings at the base of the petals. Flowers are large and free on a bush which is upright and bushy. Foliage is a pleasant dark green. Quite showy and an easy variety to time to a given date.

'Crystal Blue' (Kennett 1962). Single. The tube and sepals are white with a slight flush of pink in the tube, and the petals of the corolla are a pale violet-blue. Flowers are of medium size but freely produced. Growth is upright and willowy. Well grown this is an attractive fuchsia and with its rare colouring it adds an air of distinction to a multiple pot class.

'Curtain Call' (Munkner 1961). Double. The tube and sepals are carmine and the corolla is of rose bengal, flecked lake and crimson. The petals of the corolla are serrated. Flowers are fairly large and very free. It has that good habit of producing four blooms from each leaf axil instead of two as in the majority of fuchsias. A good all-rounder.

'Dark Secret' (Hodges 1957). Double. Tube waxy, greenish white, sepals white and reflexing to show crêpe texture, coloured underneath pale pink. Corolla deep violet-blue surrounded by small petaloids of phlox-pink. Flowers are large and fairly free on a bush which is of medium growth and upright. Attractive and admired.

'Darlene' (Scott 1952). Semi-double. The long tube is white, the sepals are white, tinted rose. Corolla purple outside with white centre, the whole changing to orchid-purple with age. Flowers are large and very free. Strong upright grower. A variety that needs very little tying and training to make a well-shaped bush. Will also make a fine standard.

'Debby' (Nessier 1952). Double. Tube, and the wide recurving sepals are of rose bengal, the large double corolla has heliotrope-blue petals which fade, with age, to cobalt-violet. Flowers are large and plentiful. Growth is strong and upright. A very nice and showy variety.

'Dorothea Flower' (Thornley 1969). Single. The tube is white and the sepals are also white but tinged with pink. The corolla is deep lavender paling to white at base, and tinged pink. Flowers profusely. Growth is willowy but quite strong and although the whole appearance of the fuchsia is delicacy, the constitution is quite robust. When introduced this cultivar was immediately popular and is common on the show bench. Named by the Dorking (Surrey) Fuchsia Group after their late President, an outstanding grower of fuchsias.

'Ecstasy' (Tiret 1948). Double. Tube and sepals rich rose, the sepals tipped green, and the corolla is rich blue with large splashings of phlox-pink. Flowers are large and free. Growth medium and rather lax. The flower is beautiful but growth is rather poor, and it takes much skilful tying to show this variety at its best. Well grown it is a winner. Suitable also for hanging baskets.

'Emile de Wildeman' (Lemoine 1905). Double. The waxy tube and sepals are rich red, the full corolla is blush-pink, shaded rose. Flowers are large and free. Growth is vigorous and upright. Very easy to grow and time for showing purposes. This variety is much better known as, and more often catalogued under, the name of 'Fascination'. Extremely popular with all fuchsia growers, whatever their aims. A worthy variety. Will also make a fine standard and pyramid.

'Fiona' (Clark 1962). Single. The tube and the long elegant sepals are a clear white. The corolla opens an attractive blue paling to a light reddish purple with age. Flowers are large and produced in good numbers. Growth is upright and very bushy. An excellent introduction.

'Flocon de Neige' (Lemoine 1884). Single. Tube and sepals scarlet, corolla white, with very distinct cerise veining on the petals. Flowers are large and very free. Growth is strong and upright. A very showy variety when in full bloom. Besides making an attractive bush, it also makes an excellent standard.

'Gay Time' (Waltz 1949). Double. Tube and sepals fuchsia-pink, corolla deep blue splashed and edged rose. Flower is of medium size and very free. Growth is vigorous and upright. An easy grower with a very lovely, showy flower. Can also be tried as a standard.

'Giant Falls' (Erikson 1957). Double. Tube pink, sepals pink with green tip, corolla creamy white. Flower is very large and produced in large numbers. Growth is quite vigorous and upright, but has not the substance in the stem to support the quantity of huge blooms this variety will give when well grown. A careful system of tying up is necessary to get the best effect of these wonderful blooms. Useful also as a trailer.

'Guinevere' (Dale 1950). Semi-double. Tube and sepals white, corolla orchid-blue. Flowers are large and free. Growth strong and upright. A variety which is very easy to grow into a well-shaped bush. Very popular both for exhibition and general greenhouse culture. Can also be grown as a standard.

'Hanora' (Travis 1957). Single. The long tube is ivory flushed with carmine. Sepals are long and open horizontally, and in colour they are carmine on the upper surface and a warm pink underneath. The rather square-looking corolla is a rich claret. The flower is largish and very free on a shrub that is upright and bushy. A worthy variety. Try also as a standard.

'Heavenly Blue' (Niederholzer 1946). Single. The long tube is pale pink, as also are the long, narrow, rolled-up sepals. The corolla is a very pale blue. The flowers are very long and very freely produced. Growth medium, upright and sturdy. A really good fuchsia having the most delicate pastel colouring. Well grown, it cannot fail to impress.

'His Excellency' (Reiter 1952). Double. The short tube and the long sepals are white. Corolla a brilliant violet ageing to orchid purple. The flower is large, and for such size, quite free. Very upright and well worth growing.

'Hollydale' (Home Gardens 1946). Double. Tube and sepals dark orchid-pink, corolla deep pink. The medium-sized flowers are very freely produced. Growth is strong and upright. Some authorities are of the opinion that this is a sport of 'Winston Churchill'. It is a good variety.

'Iced Champagne' (Jennings 1968). Single. Tube and sepals are pale pink and the corolla is described as rhodamine-pink. Extremely free flowering, the small- to medium-sized blooms are borne on a bushy upright shrub, which is short jointed and self-branching. Winner of the Jones Cup as the outstanding seedling of its year.

'Indian Maid' (Waltz 1962). Double. Tube and sepals scarlet red, the sepals being extra long, wide, and recurved. Corolla a rich shade of royal purple. The high centred bud opens with a blaze of colour which shows up well against the deep green foliage. Flowers very free. Growth willowy. With care will make a grand bush. By its nature will make an excellent basket.

'Inspiration' (Nelson 1952). Double. Tube and sepals pink, corolla pink also but of a darker shade. Flowers are of medium size but most profuse. Growth is vigorous and upright. A good pot plant that will smother itself with flower. A wonderful beginner's plant. Should also be tried in standard form.

'Jack Ackland' (Haag 1952). Single. Tube and sepals bright pink, corolla deep rose ageing to almost the same colour as the sepals. Flowers large and produced in great numbers. Growth is upright and very bushy. A well-grown bush of this variety is a mass of colour. Besides the open flowers, the large pink buds of the flowers to come assist in this mass colour effect. Very popular and worthy of that popularity. With careful training will make a good standard. Also suitable for hanging baskets.

'Jules Daloges' (Lemoine 1907). Double. The tube and sepals are scarlet, the sepals turning right back to the tube revealing the very full corolla of rich violet blue. The flower is large and free, and growth is upright and bushy. Some will describe this variety as old-fashioned, but it is none the less lovely for that.

'Kernan Robson' (Tiret 1958). Double. Tube and outside of sepals flesh pink, inside of sepals red. The large fluffy corolla is smoky red. Flowers are large and very free. Growth is strong, upright and bushy. The unusual colouring of the corolla makes this variety really stand out from all others. Quite distinct. Should also be tried as a standard.

'King's Ransom' (Schnabel 1954). Double. Tube and sepals are a clear almost transparent white, the corolla is a rich dark purple. Flowers are large and borne in profusion. Growth is strong, upright and bushy. The flower is a beautiful shape and the sepals in the bud stage are so clear as to allow the intense colouring of the enclosed corolla to show through. One of the most outstanding varieties of modern times. Should be tried also in standard form.

'Lacedemone' (Lemoine 1911). Double. Tube and sepals cerise, the full corolla is white with very faint cerise veining on the petals. Flowers are large and fairly free. Although the corolla is full it has not the untidiness of many doubles, the petals being neatly placed. Growth is upright and bushy.

'Lady Ann' (Tiret 1953). Double. The short tube and the broad spreading sepals are white faintly blushed on the underside and tipped green. The spreading corolla is formed of many curled petals, which are purplish blue surmounted by petaloids of phlox pink. Flowers are large and fairly free. Growth is strong and upright. Very elegant.

'Leonora' (Tiret 1960). Single. Tube and sepals and corolla are all a soft pink. The medium bell-shaped flowers are produced in profusion on a free-branching bushy plant. A remarkably easy variety to grow, and besides pleasing the novice is a must for the exhibitor. Will also make a perfect half standard.

'Lord Roberts' (Lemoine 1909). Single. Has a short scarlet-cerise tube, with horizontal sepals of the same colour. The corolla is very large and dark purple in colour. The habit of growth is vigorous and upright. Grown best as a tall bush. Ideal for greenhouse growing and for exhibition at the early shows.

'Lucky Strike' (Niederholzer 1943). Semi-double. Tube and sepals pale flesh-pink, corolla blue, marbled rose and pink.

253

Flowers are large and free on a shrub which is strong, upright and fast growing. Can be had in flower early in the season and it will continue in bloom for a very long time. Besides making a beautiful bush, it will also make an excellent standard.

'Madame Carnot' (Lemoine 1895). Double. Tube and sepals red, corolla blush-white streaked with carmine. Flower is large and quite free. The flower is of an extremely neat and compact appearance. Growth is medium and bushy. An old variety but still good.

'Madame Danjoux' (Salter 1843). Double. The short tube and the broad long sepals are carmine, the corolla is the palest violet-pink. Vigorous habit and very floriferous with largish flowers. Foliage a nice medium green. Excellent variety for general growing, but an ideal one for exhibition. Easy to grow, it will continue in bloom for a very long time. Good as a standard.

'Madame Jules Chrétien' (Bonnet). Double. Tube and sepals red, corolla white with heavy carmine veining on the petals. Flowers are large and very free. Growth upright and bushy. An easy fuchsia to grow.

'Mama Bluess' (Tiret 1959). Double. Tube and sepals deep rose, corolla soft blue. The flowers are very large and full and extremely free for flowers of such size. Growth is short, stocky and bushy. Grown in too lush conditions the growth is inclined to be rather lax and will not support the heavy weight of bloom. A flower which is always admired. Can be trained for hanging baskets.

'Marin Belle' (Reedström 1959). Single. Tube and sepals salmon-pink, the bell-shaped corolla is pansy-violet. The sepals reflex covering the tube, ovary and pedicle. Growth is strong and upright. Will never attain the popularity of its stable companion 'Marin Glow', but still a good fuchsia. Try also as a standard.

'Marin Glow' (Reedström 1954). Single. Tube and sepals pure waxy white, corolla imperial purple ageing slowly to magenta. Flowers are of medium size but extremely profuse. Growth is that of a strong upright bush. A fuchsia which should be in every collection. Easy to grow, easy to time, prolific in flower, and a bloom of classic fuchsia shape. Outstanding. Will also make a fine standard, and should make a good pyramid.

'Marin Monarch' (Reedström 1955). Double. The waxy tube and sepals are a glowing red. The large corolla opens to a royal purple but turns a reddish-purple with age. The striking blooms are large but are unfortunately not produced in great numbers. Growth is very strong and upright. Well timed to give maximum bloom at show date this variety could be unbeatable.

'Martin's Midnight' (Martin 1959). Double. Tube and sepals bright red, corolla very deep indigo-blue. The flowers are very large and quite freely produced. Growth is that of a strong upright bush. The rich colourings of this variety make it extremely attractive. Put it alongside one of the white varieties in the greenhouse, or on the show bench to make it stand out.

'Mauve Beauty' (Banks 1869). Double. Tube and sepals cerise, corolla a lilac-mauve. The medium-sized flowers are borne in great profusion. Growth is stocky, upright, and very free branching. This variety is extremely old but in its colour range has yet to be surpassed. One of the freest-flowering double varieties.

'Melody' (Reiter 1942). Single. Tube and sepals pale ivory-pink, corolla pale cyclamen purple. The flowers are of medium size and produced in vast numbers. Growth is strong, upright, and bushy. Besides making a first-class bush plant, this variety makes a grand standard, or an excellent pyramid. One of America's outstanding introductions. Very easy to grow and train. Extremely popular.

'Mephisto' (Reiter 1941). Single. Tube and sepals red, corolla very deep crimson. The flowers are small but very profuse, and borne in clusters. Growth is vigorous, upright and bushy. A very fast grower which will make a large plant the very first year. Very attractive when in full bloom with its myriad flowers.

'Mieke Meursing' (Hopgood 1969). Single/Semi-double. The flower has a red tube and sepals with a pale pink corolla with deeper pink veining and long stamens which are very prominent. The classic-shaped blooms are borne in great profusion on an upright and naturally bushy plant. This cultivar was immediately popular on its introduction and with the triphylla 'Billy Green' has already threatened to displace 'Snowcap' as the most common fuchsia on the show bench.

'Mission Bells' (Walker and Jones 1948). Single. Tube and sepals bright red. The corolla is bell-shaped with wavy edged petals, and is of rich purple colouring. Flowers are medium to large, and very free. Growth is strong and upright. Can be had in flower quite early in the season. A good single.

'Miss Vallejo' (Tiret 1958). Double. Tube and sepals deep pink, with a green tip to the end of the sepals, corolla deep pink, streaked with rose-pink. The flower is huge, and for such size, quite free. Growth is good, upright and bushy. The very large flowers are extremely beautiful, and the ladies adore the lovely colourings. A variety that demands admiration.

'Morning Light' (Waltz 1960). Double. Tube and base of sepals are coral-pink, while the broad upturned sepals are white on top, flushed pale pink underside. Corolla is fully double, and is deep lavender-blue, with smaller outer petals of lavender splashed pink. Corolla fades to rosy lavender with age. Flowers are large and quite free. Growth is strong and upright. Foliage is a beautiful pale green. Very new and very lovely.

'Moth Blue' (Tiret 1949). Double. Tube and sepals red, corolla lilac blue. Flowers are largish and very free. Growth is sturdy, upright and bushy. The foliage has a coppery hue to it which forms a good background to the flowers. A popular variety because it is both easy and attractive. Suitable also as a standard.

'Nautilus' (Lemoine 1901). Double. Tube and sepals cerise, corolla white faintly veined cerise. The sepals curve right back to hide the tube completely, and reveal fully the long double corolla. The flowers are large, and for such size, very free. Growth is sturdy, upright and bushy. A lovely bloom. Suitable also as a standard.

'Navajo' (Nelson 1956). Double. The tube is a pale reddish orange and the sepals are the same colour except that they have a distinct green tip. The corolla is darker, being a smoky red-orange. The blooms are large and very free. The foliage is a glossy, rich dark green which shows up the flowers well. Growth is strong, upright and bushy. The exotic colourings make this fuchsia stand out when grown with others. Should make a good standard.

'Pacific Queen' (Waltz 1963). Double. The wide phlox-pink sepals are tipped with white. The fully double corolla is a warm shade of old rose fading to a brighter shade. The big, free-blooming, easy-to-grow, upright variety with medium-sized dark green foliage. A real show fuchsia.

'Papa Bluess' (Tiret 1956). Double. Tube ivory-white, sepals pale pink, corolla deep rich violet. The flower is very large with a very full corolla, and is quite free for its size. Growth is upright and naturally bushy. A magnificent flower that is full of substance and impressive because of that.

'Party Frock' (Walker and Jones 1953). Double. The tube and the long upturned sepals, are of rose-red, the corolla is of soft lilac-lavender, flushed with flesh-pink on the outer petals. Flowers are large and quite free, and they are well displayed against the dark green foliage. Growth is strong and upright, and it will make a very large bush. Very aptly named, for the lovely pastel colours are what one might expect on a little girl's frilly party frock. With careful training will also make a good standard.

'Patty Evans' (Evans and Reeves 1936). Double. Tube and sepals waxy white to rose, corolla has white petals flushed pink. Blooms are large and produced in fair numbers. Foliage is light green and the habit of growth is fairly sturdy and upright. Although this was one of the earliest American introductions it is still one of the best. Can be trained to standard form.

'Peppermint Stick' (Garson 1951). Double. Tube and up-turned sepals carmine, but with a distinct white stripe running down the middle of each sepal, hence the name. The corolla has centre petals rich royal purple, the outer petals light carmine-rose, with purple edges. Growth is strong, upright and bushy. Flowers are large and fairly free. Can be grown also as a standard.

'Periwinkle' (Hodges 1951). Single. The tube and the long recurved sepals are pink, the long single corolla is a pale lavender-blue. Flowers can be said to be more long than large, but they are produced in profusion. Growth is upright, strong and wiry. The thin stems look as though they will not support the large number of blooms, but they are of some substance despite their appearance.

'Pharaoh' (Need 1965). Single. Tube rose bengal. Sepals are white edged with rose bengal and tipped green. Corolla opens plum purple which matures to ruby red. The medium-sized flowers are produced in profusion on an upright, self-branching bush. An excellent variety also suitable for growing as a standard.

'Pink Quartette' (Walker and Jones 1949). Semi-double. Tube and sepals deep pink, corolla a very pale pink. Very aptly named, for the corolla is composed of four distinct rolls of petals. Flower is large and very free. Vigorous grower with a strong upright habit. An extremely showy variety which is really worth growing. Should also be tried in standard form.

'Pin-wheel' (Waltz 1958). Double. A large, flat, double upright in pastel shades. Flower resembles a pin-wheel in form. Double corolla a lovely shade of violet, has folded petals which open wide in the centre. Short tube and buds are flushed pink. The pink sepals are broad, pointed, and turn back slightly at tips but do not lose their flat appearance. An excellent intro-duction.

'Pio Pico' (Tiret 1955). Double. Tube and sepals very pale pink, corolla violet-purple shaded wine, and splashed with pink.

Flower is immense, and for such size, very free. Growth is upright and bushy but support is needed when the heavy blooms are there. An eye-catching variety which responds to good training. Suitable also for hanging baskets.

'President S. Wilson' (Thorne 1969). Single. The tube is carmine, and the sepals are carmine at the base turning to pale pink, and each is tipped green. Corolla is rosy carmine. The flower is very large for a single and they are borne in great numbers over a very long period. Growth starts off very upright but turns to trailer with the weight of the numerous blooms and buds. With training this cultivar will make a fine bush, espalier, basket, or weeping standard plant. The author is proud to have such a fine cultivar named after him.

'Puget Sound' (Hodges 1949). Double. Tube and sepals rosy red, corolla clear white with petaloids overlaid in light pink. The corolla is very large and fluffy. Growth is that of an upright spreading bush. Flowers are large and very freely produced. A good variety which, when well grown, is difficult to better.

'Purple Heart' (Walker and Jones 1950). Double. Tube and sepals crimson, corolla has outer petals of deep rose-red, and inner petals of rich violet-purple. Flower is huge although not very free. Growth is strong and upright. The flowers are among the largest that have come from America, but this variety can be temperamental, for too violent fluctuations in temperature result in the flower refusing to open. One for those who prefer size to quantity.

'Radiance' (Reiter 1946). Semi-double. Tube and sepals crimson, corolla crimson at base and edges, Tyrian rose in centre. The petals are rather irregular. The medium-sized flowers are very free. Growth is upright and bushy. The flowers have a charm of their own, and on a well-grown plant with a quantity of bloom this shrub can be outstanding.

'Raindrops' (Evans and Reeves 1954). Single. Tube and sepals clear white, corolla rich rose-red. Flowers are large and quite profuse. Flowers show up well against the rich green foliage. Growth is strong but rather lax, and careful tying is necessary to get a well-shaped bush. Very attractive. Suitable also for hanging baskets.

'Rolla' (Lemoine 1913). Double. Short tube and sepals very pale pink, corolla pure white. Very floriferous with a vigorous habit of growth. Flowers large and about 3 inches long. Foliage light green. A very attractive fuchsia which besides being ideal for growing in bush form will also make a very pretty standard.

'Royal Velvet' (Waltz 1962). Double. Tube and upturned sepals crimson-red. The large, double, ruffled corolla opens a luminous deep purple, and as it matures the petals flare open displaying contrasting crimson centre and the colour changes to rosy purple. Long crimson pistils add to the beauty of this flower. Extremely free flowering for so large a flower. Excellent upright habit of growth. Besides making a good bush will also make an excellent standard and a worthy basket.

'Rufus the Red' (Nelson 1952). Single. Tube, sepals and corolla all bright turkey-red. The medium-sized flowers are produced in great numbers and over an extremely long season. A very fast-growing variety, with strong upright growth. Will make a fine bush the first year. Worth a place in any collection. Will also make a fine standard.

'San Jacinto' (Evans and Reeves 1952). Double. Tube and sepals shell-pink, corolla lilac but fading to lilac-pink with age. The medium-sized flowers are extremely freely produced, on a shrub that is upright and very bushy. The flower is really beautiful and for a double, most tidy. A variety that is easy to grow and almost impossible to fault. Try also in standard form.

'San Leandro' (Brand 1949). Double. Tube and sepals magenta, corolla a mixture of magenta and vermilion. Very large flowers and for their size very freely produced. Growth is vigorous and upright. This is a variety which only grows well if allowed to reach a good size, by making a final potting into a 10-inch pot. Not free branching, so looks a little absurd in the smaller pots. A fine fuchsia which will make an excellent specimen plant.

'Sherl Ann' (Peterson 1958). Double. The tube and the thick sepals are bright red. The corolla is a full double white, the centre of each petal flushed with red veining. A most prolific bloomer, the flower has excellent form. Growth is strong and upright. Suitable for standards.

'Shy Lady' (Waltz 1955). Double. Tube and sepals ivory-white, corolla opens almost white, and a very definite pink blush comes into the petals as the flower ages. The flowers are large and extremely free. Growth is upright, free branching and bushy. This is really a beautiful fuchsia, and must become more popular as growers get to know it. Easy to grow, and admired by everybody Very cleverly named.

'Sleigh Bells' (Schnabel 1954). Single. Tube, sepals and corolla all pure white. The corolla is bell-shaped. The medium-sized flowers are very free. Growth is upright and bushy. This is perhaps the largest single white fuchsia to date, and like all the white fuchsias it will bruise and water-mark very easily.

'So Big' (Waltz 1955). Double. Tube and sepals palest pink, corolla creamy white. The flowers are huge, but there are not many of them. Growth is weak and much tying is necessary to get a good-looking bush. This is one of the largest flowers in the fuchsia world, but it lacks the growth to support it. For exhibition purposes the grower must pay particular attention to timing, so that the maximum bloom is there for the judges on show day. Only the grower who is fascinated by size will grow this variety.

'Spanish Rhapsody' (Colville 1965). Double. Tube and sepals are red. The corolla is a darkish blue with purple shading and has a satin-like sheen which is quite noticeable. The flower is large and full and very freely produced. The growth is upright and bushy. This cultivar won the Jones Cup in 1965 for being the outstanding seedling of that year.

'Stella Marina' (Schnabel 1951). Semi-double. Tube and sepals pale crimson, corolla violet-blue irregularly splashed crimson and white. Flower large and very free. Growth is strong, upright and very bushy. A variety that will appreciate a cool shady corner of the greenhouse where it will give of its best. Worth growing if a corner to suit it can be found. Will also make a charming standard.

'Sunset' (Niederholzer 1938). Single. Tube and sepals pale pink, corolla orange-cerise. The medium-sized flowers are quite distinctive with their wide-open corollas. Extremely prolific in flower, this variety will flower continuously from early in the season until the late autumn. Very easy to grow and train. Growth is upright and naturally bushy. Can be trained to standard form, or for hanging baskets.

'Swanley Gem' (Cannell 1900). Single. Tube and sepals rich scarlet, corolla violet. The four petals of the corolla open flat, making a perfect circle. The flowers are of medium size and most profuse. Growth is upright and quite naturally bushy. A very popular variety and very worthy of that popularity. Always admired on the show bench or in the greenhouse.

'Sweet Leilani' (Tiret 1957). Double. The short tube and the sepals are flesh pink, the corolla can only be described as a smoky blue. The petals of the corolla are very wide and spreading. The flower is large and quite free. Growth is quite strong, upright and bushy. A variety that is always admired.

'Symphony' (Niederholzer 1944). Single. Tube and sepals pale pink, corolla bluish mauve. The flower is large with a large spreading corolla. The flowers are quite freely produced on a bush which is strong and upright. A good fuchsia, and ideal for the show class which calls for a single variety.

'Tangerine' (Tiret 1949). Single. An interesting hybrid from *F. cordifolia* having a flesh-coloured tube, distinct green sepals and a corolla which opens as orange and gradually ages to rose. The flowers are slim and long and are quite freely produced. Growth is strong, sturdy and upright. A lovely variety which puts the unusual into an exhibit or into the greenhouse collection.

'Tanya Bridger' (Bridger 1958). Double. Tube and sepals white, corolla pure lavender-blue. Flowers are of medium size and most profuse. Growth is upright and bushy. Will often grow better in a sheltered spot in the garden rather than in the greenhouse, as too 'lush' conditions tend to make growth rather soft and the shrub suffers in consequence. Well grown, this is a lovely variety.

'Tennessee Waltz' (Walker and Jones 1951). Double. The tube and the upcurved sepals are rose madder. The squarish petals which form the corolla are lilac-lavender, splashed with rose. Flowers are medium to large and extremely free. Growth is strong, upright and bushy. A very easy variety to grow and strongly recommended to all growers. This is probably one of the most popular varieties grown today. Ideal also as a standard, and should make an excellent pyramid.

'The Aristocrat' (Waltz 1953). Double. The tube and the wide upturned sepals are of pale rose. The very full corolla has petals which are creamy white, veined rose at base, and which also have a very serrated edge. The flowers are large and fairly free. Growth is sturdy and upright. A real aristocrat.

'Theroigne de Merricourt' (Lemoine 1903). Double. The tube and sepals are scarlet, the latter being broad and horizontal. The corolla is creamy white, slightly veined red under the sepals. It has a vigorous habit of growth and is fairly free, the petals of the corolla being so placed as to give a very full appearance on the 4½-inch-long flowers. The foliage of dark sage-green provides an excellent background for the flowers. Will make a large bush.

'Toby Bridger' (Bridger 1958). Double. Tube and sepals bright pink, corolla pale pink. The corolla has petals of very neat placement on flowers that are large and profuse. Growth is sturdy and compact. This variety is extremely popular and is thought by many to be the finest of the modern British raised varieties.

'Torch' (Munkner 1963). Double. The waxy tube has very broad sepals which are shiny pink outside, heavily flushed salmon inside. Double corolla of two distinct colours. Purplish red in the centre, with orange-salmon outer petals. Growth is tall and upright. A very showy fuchsia.

'Treasure' (Niederholzer 1946). Double. The tube is ivory, sepals are pale rose, and the corolla is violet-blue. The corolla is large and fluffy in appearance. Growth is upright, branching and bushy. A good attractive fuchsia.

'Tristesse' (Blackwell 1965). Double. The tube is a pale rose-pink, and the sepals are a deeper pink, tipped green. The corolla is a pale lilac-blue. The medium-sized blooms are extremely free on a low, self-branching bushy plant. Grows naturally to a good shape and needs but little training. An easy plant to grow but always admired.

'Ultramar' (Reiter 1956). Double. Tube and the long, broad, recurved sepals are a creamy white. The flowers are very double in appearance, and the petals open into a formal globular corolla of a delicate pastel 'grey-blue'. Where the corolla and sepals meet are a number of white-streaked petaloids. Growth is strong and upright. A handsome fuchsia.

'Vagabond' (Schnabel 1953). Double. The short tube, and the long broad upturned sepals are brilliant carmine, the many-petalled corolla is magenta with some of the outer petals splashed with carmine. The large flowers are profuse on a shrub that is upright, but in need of careful tying to support the wealth of bloom. Can be used in hanging baskets.

'Vanity Fair' (Schnabel-Paskesonn 1962). Double. Thick green-white tube and sepals. Large pale pink, heavily petalled globular double corolla with serrated edges resembling a carnation. Flowers very free. Growth upright bush.

'Vicky Putley' (Putley 1964). Single. Tube and sepals white-flushed carmine. Corolla crimson and rather shortish. A very free-flowering variety. Growth is rather lax and requires plenty of ties to effect a good bush shape. Could be tried as espalier and basket.

'Violet Gem' (Niederholzer and Waltz 1949). Semi-double. Tube and sepals carmine, the spreading corolla is deep violet. Flowers are large and very free. Growth is upright and bushy. It is difficult to fault this variety as it is so easy to grow.

'Violet Rosette' (Knechler 1963). Double. The tube and the short wide sepals are bright carmine with the sepals growing straight back. Large fully double corolla is a deep violet with a touch of red at the base of the petals. Very free flowering. Foliage a bright green which is the ideal foil for the flower. Growth upright.

'Voo-doo' (Tiret 1952). Double. Tube and sepals very dark red, corolla very dark purple-violet. Flower is quite large and freely produced. Growth is upright and bushy. As the name indicates, this is a rather sombre-looking variety. Grown in the greenhouse, or shown on the exhibition table next to one of the white varieties, this fuchsia will look quietly charming while showing off the brilliance of the other plant's white flowers. Should make a good standard.

'Whirlaway' (Waltz 1961). Semi-double. Tube and sepals and corolla are all milky white which develops a delicate blush tint as flower matures. The tube is short and the sepals extra long. The long flowers are extremely free. Growth is willowy.

'White Bouquet' (Waltz 1959). Double. Tube, sepals and corolla all white except for a touch of pink at the junction of the tube and the sepals. The blooms are large and very freely produced. Growth is upright and bushy. A beautiful variety with long stamens and pistil in an almost transparent pink showing up well against the white corolla.

'Wood Violet' (Schmidt 1946). Double. Tube and sepals dark red, corolla violet-blue. The medium-sized flowers are produced in profusion. Growth is sturdy, upright and bushy. Makes an excellent bush plant.

'Yuletide' (Tiret 1948). Double. Tube and sepals crimson, the large corolla is creamy white. The bloom is large and quite freely produced. Growth is strong, upright and bushy. A very showy variety. Can be grown to standard form.

CHAPTER XXII

Recent Introductions

Every year dozens of new varieties of fuchsia are raised by both the British and the American hybridizers. The grower must guard against believing that the new varieties are improvements on the existing ones which have proved themselves over the years. Again, many that are listed as 'new' are so similar in many ways to varieties that have been grown for years past that their introduction is not really merited.

Gardening, to be enjoyed, must be treated as an adventure, and it is essential that whatever is put on to the market must be tried before a verdict is passed on it.

The following varieties are many of the recent introductions which have yet to prove their worth. The adventurous grower will want to try some of them and may well find among them one, two or more varieties which will please him, or her, and may in time become established favourites in the gardening world.

The descriptions given are, in most cases, those given by the raisers to their own creations.

'A. M. Larwick' (Cox 1968). Single. Tube and sepals are rich carmine, and the corolla is a purplish mauve, veined with carmine. The flower is large for a single, and is produced in large numbers. Growth is upright and bushy. Ideal as a summer bedder, but will also make an excellent standard. An introduction from New Zealand.

'Bishops Bells' (Gadsby 1970). Single. A well-shaped large-flowering single seedling from 'Caroline'. The short tube and broad sepals are deep crimson. The corolla is bell-shaped and opens petunia-purple and matures to deep orchid-purple.

'Bora Bora' (Tiret 1966). Double. The tube is white and the sepals are white on top, pink underneath, and tipped green. The corolla is purplish blue. The flowers are fairly free and are produced on a rather lax-growing but free-branching shrub.

'Cliffs Hardy' (Gadsby 1970). Single. A single hardy variety, tube and sepals crimson, corolla campanula-violet. Another seedling from the same parents as 'Upward Look' and when grown outside the flowers are just as upright, though the sepals are long and spiky.

'Corsair' (Kennett 1965). Double. The tube and sepals are waxy white. The large double corolla opens a sky-blue and then fades to light purple. Centre is white and white runs up centre of each petal. The surrounding petaloids are white with purple marbling. Free flowering on bushy plant. Striking contrast of colours.

'Derby Belle' (Gadsby 1970). Single. Another seedling from 'Caroline', with a bluer corolla and very long thin sepals. Tube and sepals are white, flushed pink. The bell-shaped corolla opens light Bishop's-violet and matures to light orchid-purple.

'Diana' (Kennett 1967). Double. Sepals white, flushed with pink. The large corolla is a light marbled lavender which fades to a bright old rose. A lax, trailing habit of growth. Foliage is light green.

'Diana Wills' (Gadsby 1971). Double. A bushy plant with large double flowers very freely produced. The tube and sepals are white with green tips to the sepals. The corolla opens deep petunia-purple, with rhodamine-purple at the base, and matures to rhodamine-purple all over with pink on the outer petals.

'Edith Summerville' (Miller 1970). Single. A single with long narrow tube of pale crimson and with sepals of the same colour. The petals of the corolla are pale petunia-purple. Has a bell-shaped corolla of classic form. Flowers of good size and very free.

'Ethel' (Martin 1967). Semi-double. The large flower has a corolla which opens with a lavender centre with pink petaloids on the outside. As it matures it spreads out and fades to pale orchid. Sepals are white with green tips and pale pink on underside. Growth upright.

'Fiery Spider' (Munkner 1960). Single. Corolla crimson with decided orange flush. The long thin tube is deep carmine. Long narrow sepals are pale salmon, tipped light green. Good strong cultivar with long blooming season. Makes an ideal basket.

'Flame of Bath' (Colville 1971). Double. Tube and sepals are orange-red, corolla flame-red, flushed orange. Flowers are large and free. Growth is upright and bushy.

'Flashlight' (Gadsby 1971). Single. A hardy seedling the result of a cross between 'Flash' and 'Magellanica Alba', and which seems to have got rid of the pale lilac of all the earlier Mag. Alba seedlings. The flowers are about twice as big as Mag. Alba and a pale phlox-pink in colour.

'Frosty Bell' (Gadsby 1970). Single. The sepals and tube are pale pink and the corolla is pure white. The medium-sized flowers are bell-shaped and very free. They are set off well against the long, pointed, shiny dark green foliage. Growth is upright and bushy.

'Glory of Bath' (Colville 1971). Double. Tube and sepals are bright red. Corolla is mauve veined turkey-red. Flowers are large and free. Growth willowy.

'Goldcrest' (Thorne 1968). Single. Tube is light pink. Sepals are light pink, paling at the tip and turning to green. Corolla is lavender veined pink. A large flower very freely produced. Growth is lax but very free branching. The delicate colouring of the bloom stands out well against a foliage which is golden but which gradually turns to light green with age.

'Lightning' (Gorman 1970). Double. The long tube and sepals are ivory coloured and the sepals themselves are tipped with green. The corolla is a striking orange-red and is also quite long. The handsome flowers are borne quite freely, and they stand out against the dark green foliage. Growth is upright and bushy but can be made to trail.

'Nightingale' (Waltz 1960). Double. The short tube and sepals are white, flushed pink. The frilly double corolla has a centre of deep purple which matures to a bright magenta, and this is surrounded with outer petals of pink, white, and coral tones. Blooms are of a heavy texture and long lasting. Growth is upright but tends to trail when the heavy blooms are out.

'Purple Emperor' (Endicott 1970). Semi-double. Both tube and sepals are a rich crimson and the corolla opens as violet but matures to purple. Growth is strong, bushy and upright. A most prolific bloomer which is easy to grow.

'Rahnee' (Colville 1966). Double. The tube and sepals are a deep pink and the corolla is lighter but still a definite pink. The flowers are described as medium to large and are freely produced. Growth is upright and bushy.

'Razzle-Dazzle' (Martin 1965) Double. The tube and sepals are pink and the sepals are distinctive in that they curl up tightly. The corolla is dark purple with an almost blackish-purple edge. The corolla never really opens out and remains tight and compact. Growth is of a rather lax habit.

'Snowdrift' (Kennett 1966). Semi-double. Both tube and sepals are white. The large flaring corolla is also white, but is tinged slightly with pink. Foliage is a glossy dark green. Growth is of a low bushy habit.

'Stanley Cash' (Pennisi 1970). Double. The short tube is white as also are the very short sepals, except that they have green tips. The corolla is large and of squarish appearance and in colour a deep royal purple. Growth starts off upright but is soon weighed down by the blooms.

'Strawberry Delight' (Gadsby 1970). Double. The waxy tube and sepals are crimson although the under portion of the sepals is carmine. The corolla is white but flushed with carmine well down into the outer petaloids. Foliage is a golden bronze. Growth is medium and upright. This variety is an early bloomer, and is quite hardy. Small flowers for a double.

'The Marvel' (Vicarage Farm Nurseries 1969). Single. The tube is white and the sepals are also white, but are tipped with green. The neat corolla is pink, but an unusual pink, having a lavender sheen on it. A strong growing, upright and bushy cultivar, which is very free flowering and is ideal either as a greenhouse pot plant or growing for bedding out in the summer. The flowers are very solid and last well on the plant.

'White Bride' (Gadsby 1970). Double. The tube is white while the reflexing sepals are white tinted with pink and with distinct green tips. The long double corolla is pure white. Leaves are rather large and roundish. Growth is upright and is that of a medium bush.

'Wingroves Mammoth' (Wingrove 1968). Double. Both tube and sepals are turkey-red. The huge fluffy corolla is white, veined and heavily splashed with carmine. Growth is upright until the arrival of the very large blooms which tend to weigh the branches down, making tying a necessity.

'Wings of Song' (Blackwell 1968). Double. Both tube and sepals are a bright rose-pink. The medium-sized corolla is a lavender-pink with a more definite pink veining. This is a vigorous growing, free-branching, and very floriferous fuchsia. Its natural habit is cascade which makes it an ideal basket plant. However, if one wishes to take the trouble of much tying up, it will make an attractive bush.

APPENDIX I

British Nurseries that Specialize in the Fuchsia

The following nurseries are those which have for many years specialized in selling, and furthering the cult of, the fuchsia.

B. & H. M. Baker,
Bourne Brook Nurseries,
Greenstead Green,
Halstead,
Essex.

Ben Lomond Nursery,
Balmaha,
Near Glasgow,
Scotland.

J. R. Blackwell,
Kingsdown Nurseries,
35 Kingsdown Road,
Kingsdown, Swindon.

H. A. Brown,
20 Chingford Mount Road,
South Chingford,
London, E.4.

J. W. Cole & Son,
16 Holditch Street,
Peterborough.

W. Crockett,
Uplands Nurseries,
Meadow Road,
Great Gransden,
Sandy, Beds.

'Frenchakons'
157 Coventry Road,
Burbage,
Hinkley, Leicestershire.

C. J. Castle & Son,
Hillcrest Nurseries,
Church Bank,
Goostrey, Crewe,
CW4 8PH.

J. W. Jackson & Son,
The Grove Nurseries,
Grove Lane,
Timperley,
Altrincham, Cheshire.

271

Kelrose Garden Centre,
Little Warley,
Near Brentwood, Essex.

Leavesden Nursery,
Horseshoe Lane,
Garston,
Watford, Herts.

C. S. Lockyer,
74 Cock Road,
Kingswood,
Bristol, BS15 2SG.

The Marquis of Bath's
  Estate Garden,
Longleat,
Warminster, Wilts.

V. T. Nuttall,
Markham Grange Nurseries,
Brodsworth,
Doncaster.

R. & J. Pacey,
Stathern,
Melton Mowbray,
Leicestershire.

Ravensdale Nurseries,
Wykern,
Coventry, CV2 5GQ.

G. H. Roe,
Gable Nurseries,
Radcliffe-on-Trent,
Nottingham, NG12 2HT.

Vicarage Farm Nurseries,
256 Great West Road,
Hounslow,
Middlesex.

West End Nurseries,
16 Derby Road,
Luton, Beds.

Westwood Nurseries,
Gorton Brook,
Brindle,
Near Chorley,
Lancs.

E. Wills,
Fuchsia Nursery,
Chapel Lane,
West Wittering,
Chichester.

APPENDIX II

Fuchsia Nurseries of the U.S.A. and the British Commonwealth

The majority of the new varieties of fuchsia which are put on to the market each year are imported from the U.S.A. British hybridizers are catching up, and are producing some very worthy material, but they are greatly outnumbered and are likely to remain so for many years to come.

Not all American varieties are worthy of introduction into this country for various reasons. In the first place many are bred to suit the conditions that prevail on the west coast of that country, and secondly the type of fuchsia preferred by the American public and the British growers appears to differ. The British are known to go more for the bush and upright types while the Americans, judging by the glut of trailing varieties that they produce each year, have a definite leaning towards the type that is more suited to hanging baskets in this country.

The greater number of the imported varieties are received in this country in the form of unrooted cuttings. Wrapped in polythene and sent over by air, they are hardly upset in any way and they take easily to normal propagation. Rooted plants can be exchanged between America and Great Britain but all soil must be washed off before dispatch can be made.

The import regulations to America are far more severe than they are to this country. To obtain an export licence means that the grower's plants must be inspected every year by an official of the Ministry of Agriculture and Fisheries. Every three years samples of the growing compost are inspected to ensure that there is no trace of the presence of golden nematode, *Heterodera rostochiensis*—eelworm.

With the complex regulations and the inconvenience of dollar exchange it is really impracticable for the amateur grower to undertake the import or export of fuchsias. The British and the American nurserymen are constantly on the lookout for the best varieties from whichever part of the world is raising new fuchsias. The furtherance of the flower and the satisfaction given to their customers are essential to the smooth running of their businesses, and for this reason the grower can rely on his nurseryman to help and advise, and to see that the right varieties are made available.

The friendship and understanding that exist between fuchsia growers throughout the world can be seen by the international nature of their different lists of members. Because of the international interest in this flower, and since we have already listed the British nurseries, it is only right that people should know of some of the American nurseries that have done so much to ensure the continued popularity of the fuchsia. Here then is a list of such nurseries:

Abel's Camelia & Fuchsia
  Gardens,
17291 Taylor Lane,
Occidental,
California 95465.

Adkins Flowers, Fuchsias
  & Geraniums,
Rt. 2, Box 400,
Coos Bay,
Oregon 97420.

Andreason's Greenhouse,
3381 Old Elmira Road,
Eugene,
Oregon 97402.

Angeli's Nursery,
484 E. 14th Street,
San Leandro,
California 94577.

Ann's Gardens,
Rt. 3, Box 1879,
Coos Bay,
Oregon 97420.

Aumack Gardens,
Rt. 1, Box 1050 (south end
  of Ocean Drive)
Fort Bragg,
California 95437.

Barrett's Wake Robin Farm
  Nursery,
594 Wake Robin Lane,
Corvalils,
Oregon 97330.

Barringer Greenhouses,
11560 S.E. Stark,
Portland,
Oregon 97216.

274

Barton Nursery,
Hepburn Road,
Glen Eden,
Auckland,
New Zealand.

Berkeley Horticultural
  Nursery,
1310 McGee Avenue,
Berkeley,
California 94703.

Blue Horizon Nursery,
Coast Highway 1,
Gualala,
California 95445.

Carlson's Nursery,
Rt. 2, Box 451,
Poulsbo,
Washington 98370.

Copley Gardens,
2805 Chemawa Road, N.E.,
Salem,
Oregon 97303.

Craig's Fuchsia Gardens,
13128 Occidental Road,
Sebastopol,
California 95472.

Dayka Nursery,
1643 Calle Canon,
Santa Barbara,
California 93105.

Dobb's Greenhouse,
Route 10, Box 148,
Olympia,
Washington 98501.

Encinal Nursery,
2057 Encinal Avenue,
Alameda,
California 94501.

Flora Vista Gardens,
4121 Rosedale Avenue,
Victoria, B.C.,
Canada.

Florence's Fuchsias,
Rt. 3, Box 801,
Coos Bay,
Oregon 97420.

Foothill Fuchsia Farm,
261 Foothill Blvd,
San Luis Obispo,
California 93401.

Frances Fuchsia Gardens,
8761 N. Delaware Avenue,
Portland,
Oregon 97217.

Friendly Fuchsia Farm,
$752\frac{1}{2}$ Alphonso Avenue,
San Luis Obispo,
California 93401.

Fuchsia Forest,
9234 'E' Street,
Oakland,
California 94603.

Fuchsia Nursery,
14091 Old Cazadero Road,
Guernewood Park,
California 95446.

Gardenside Fuchsias,
2828 S.E. 157th Avenue,
Portland,
Oregon 97236.

Gentry's Fuchsia Garden,
412 Fremont Avenue,
Pacifica,
California 94044.

Guasco Fuchsia Garden,
515 Aspen Road, Box 222,
Bolinas,
California.

Hansen's Nursery & Florist,
3418 Union Street,
Eureka,
California 95501.

Hastings Fuchsia Gardens,
Mrs. Robert W. Hastings,
302 S. Nardo Avenue,
Solana Beach,
California 92075.

Hensel Hardware & Nursery,
884 9th Street,
Arcata,
California 95521.

Herman Nursery,
8000 Allison Avenue,
Dayton,
Ohio 45415.

Hidden Springs Nursery,
Hector Black,
Rt. 3, Rockmart,
Georgia 30153.

Highway 20 Fuchsia Garden,
Half mile east of Highway 1
    on Highway 20,
Fort Bragg,
California 95437.

Hilltop Fuchsia Gardens,
3972 Chanate Road,
Santa Rosa,
California 95404.

Island Gardens,
The Gessfords,
Rt. 1, Box 117½,
Portland,
Oregon 97231.

Ivy Banks Nursery,
Donald E. Tooke,
Rt. 4, Box 690,
Boring,
Oregon 97009.

K & F. Gardens,
35-37 Bay Street,
Osterville,
Massachusetts 02655.

Karshner Road Greenhouse,
Leslie B. Stinchcombe,
1019 Karshner Road,
Puyallup,
Washington 98371.

Kent Fuchsia Nursery,
Alan R. Kent,
60 Lucerne Crescent,
Alphington 3078,
Victoria,
Australia.

Kercsak's Fuchsia Garden,
Steve Kercsak,
103 Vista del Mar,
Pismo Beach (Shell Beach),
California 93449.

Lila's Nursery,
4 Altena Street,
San Rafael,
California 94901.

Lindrea Fuchsia Nursery,
(Regd) Kenneth C.
    McLister, Prop.,
11 Padua Court,
Glen Waverly, 3150,
Victoria,
Australia.

Manhattan Garden Supply,
305 N. Sepulveda Blvd,
Manhattan Beach,
California 90266.

McCormick's Fuchsialand
    Nursery,
Frank L. McCormick,
12419 South Laurel Avenue,
Whittier,
California 90605.

Merry Gardens,
Camden,
Maine 04843.

Milani Nursery,
10841 San Pablo Avenue,
El Cerrito,
California 94530.
Phone 237–7644.

Millars Fuchsia Gardens,
Box 102, Pacific City,
Oregon 97135.

Mission Trails Fuchsia
    Gardens,
Mr. & Mrs. W. D.
    McPheeters,
176 Crazy Horse Road,
Salinas,
California 93901.

Nix Nursery,
4707 Cherryvale Avenue,
Soquel,
California 95073.

Ocean Bluff Fuchsia
    Gardens,
Rt. 1, Box 527,
Pacific Way,
Fort Bragg,
California 95437.

Prentice Fuchsia Gardens,
3657 Highway 101 South,
Coos Bay,
Oregon 97420.

Redberg Nursery,
Mrs. Edna Redberg, Prop.,
Oretown Rt., Box 14,
Cloverdale,
Oregon 97112.

Reimers, Ben,
1113 Lincoln Avenue,
Alameda,
California 94501.

Rheas Fuchsias,
P.O. Box 43,
Burton,
Washington 98013.

Ruth's Greenhouse,
1407 N. Lilly Road,
Olympia,
Washington 98501.

Sacramento Redwood
  Shop & Nursery,
4501 13th Avenue,
Sacramento,
California 95820.

Schmidt Nursery,
355 Lambert Avenue,
Palo Alto,
California 94306.

Sherry's Fuchsia Gardens,
Rt. 2, Box 350,
Raymond,
Washington 98577.

Sherwood Forest Nursery,
2623 Harris Street,
Eureka,
California 95501.

E. C. Smith Dandenong
  Nurseries,
20 Maurice Street,
Dandenong,
Victoria 3175,
Australia.

Soo Yun's Fuchsia Gardens,
8306 Bodega Avenue,
Sebastopol,
California 95472.

Sophia's Fuchsia Gardens,
324 Kieth Avenue,
Pacifica,
California 94044.

Stafford Nursery,
18931 86th Avenue, Box 42,
Port Kells,
B.C., Canada.

Strozyk Gardens,
P.O. Box 302,
Frances,
Washington 98543.

Stubbs Fuchsia Nursery,
770 Oceanview,
Leucadia,
California 92024.

Sunrise Fuchsia Garden,
14747 S.E. 39th,
Bellevue,
Washington 98004.

Talnadge's Fern Gardens,
354 G. Street,
Chula Vista,
California 92010.

The Flower Farm,
Rt. 3, Box 587,
Junction City,
Oregon 97448.

Vicente Begonia Gardens,
1100 Vicente Street,
San Francisco,
California 94116.

Waltz Gardens,
148 Shady Lane,
Ross, Marin County,
California 94957.

Warren's Nursery,
2200 Fifth Street,
Berkeley,
California 94710.

Westover Greenhouse,
1317 North 175th Street,
Seattle,
Washington 98133.

Westport Nursery & Fuchsia
   Garden,
Alfred A. Spain,
P.O. Box 345,
Westport,
Washington 98595.

Whalebone Greenhouse,
Rt. 2, Box 351,
Raymond,
Washington 98577.

White's Fuchsia Garden,
2974 32nd Street,
Sacramento,
California 95817.

Williamson's Greenhouse,
16266–11th N.E
Seattle,
Washington 98155.

Willow Creek Nursery,
1801 Bodega Avenue,
Petaluma,
California 94952.

Wishing Well Nursery,
Mrs. Priscilla Phelps,
306 Bohemian HWY,
Sebastopol,
California 95472.

Index to Species and Varieties

281

288